The MACHINIST'S **SECOND** BEDSIDE READER
and
The Bullseye Mixture

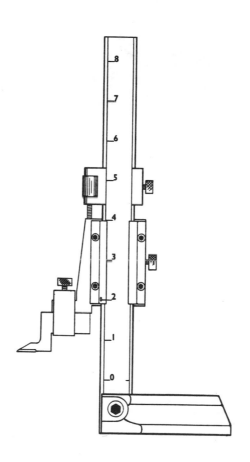

WARNING and DISCLAIMER

Metalworking is an inherently dangerous activity. Both hand- and power-operated tools can inflict serious and/or permanent damage and/or fatal injury. The information in this book is provided "As Is", without warranty. Warnings re potential dangers associated with anything described herein are not exhaustive, and cannot cover all eventualities. Neither the author, publisher, or distributor shall have any liability to any person or entity with respect to any injury, loss or damage caused or alleged to be caused directly or indirectly by information or instruction contained in this book. It is YOUR responsibility to KNOW and USE safe working practices and procedures. If you don't know how to do a job safely, don't try to do it until you find out how to do it safely in your own situation.

The MACHINIST'S **SECOND** BEDSIDE READER
and
The Bullseye Mixture

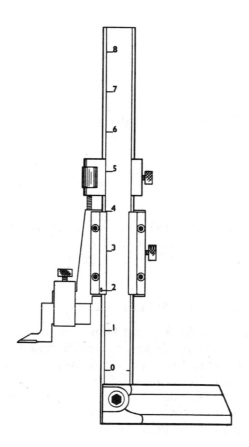

written and published
by

Guy Lautard
2570 Rosebery Avenue
West Vancouver, B.C.
Canada V7V 2Z9
(604) 922-4909

www.lautard.com

Published by Guy Lautard; printed in Canada

Second edition, 8thprinting; February 2001

Canadian Cataloging in Publication Data:

Lautard, Guy, 1946-
    The machinist's second bedside reader

ISBN 0-9690980-3-0

I. Machine-shop practice. 2. Machinists'
tools. I. Title.
TJ1160.L39 1988   670.42'3   C88-091406-8

Material from *American Machinist* Magazine, 1931, has been used with the permission of the publisher, McGraw-Hill, holder of the copyright to said material. Author, title, and copyright date (date of publication in *American Machinist)* are as indicated with the item.

BOOK DETAILS

For those who might be interested in such details, this book was prepared on an ITT "286 ATW" computer c/w 40 MB hard drive, dual floppy drives, and color monitor. Final output was done on a HP LaserJet printer, the typeface used being primarily 10 point Times Roman. Final paste up and fitting of drawings to text was done the old-fashioned way, by the author, who also sweated blood over the illustrations. The latter were done as pencil drawings and then photocopied to get good black lines. Lettering for the drawings was typed, photo-reduced to 75% of full size, and then pasted into place on the drawings.

DISCLAIMER

The list of suppliers at the end of this book is by no means complete; there would be hundreds of tool stores, and many mail order tool suppliers, of which I have never heard. The ones mentioned are simply ones of which I *have* heard. No endorsement of one company over another is implied by order of listing, or by the fact of being listed or not listed. Furthermore, mention in this book of a specific product, manufacturer, or company is not intended as an endorsement, nor should it be construed as such.

TABLE OF CONTENTS

Part Four - Miscellaneous Hints, Methods, and other Goodies

Part Five - Useful Info and Fun Projects

Part Six - Anecdotes, Minor Items and The Bullseye Mixture

Appendix A - Catch-up from TMBR, etc.

Appendix B - List of Suppliers
(pages 200 & 201)

ACKNOWLEDGEMENTS

I would like to thank all who sent me ideas for inclusion in this book. Many were used, some were not. I have tried to give credit with anything that was used, but if you sent me something, and see it herein with credit given to somebody else, don't be peeved: several people sent similar ideas.

I would particularly like to mention - and thank - two individuals whose names occur here and there throughout this book: "Mac" Mackintosh, and Tim Smith.

Mac is a retired tool & die maker, instrument maker and more, who lives in Cornwall, England. He wrote me numerous helpful letters when I queried him about one thing or another regarding the Pantograph Engraving Machine and/or other items which you will find herein. His replies were always prompt, patient, courteous and complete. I like that.

Tim earns his daily bread as a tool & die designer in Toledo, Ohio. He is also a true craftsman, in terms of the quality of work he turns out in his basement workshop, and a real do-er. Tim is a "kindred spirit" to me in many ways - he's fussy about his tools, and he likes everything done right and lookin' good. I have come to regard Tim as a good friend, and I think you will like some of the things herein that came from his hand.

In addition, Bill Fenton was always willing to take a few minutes (or more) on the phone to discuss anything I wanted to run past him. Bob Haralson was similarly willing to dig into his vast store of knowledge and experience to answer questions, or tell me about something he thought would be of interest - and it always was.

So here's to Mac and Tim and Bill and Bob.

A Special Acknowledgement

is due my wife Margaret for her unstinting help and encouragement. To me, she epitomizes "The Good Wife" described in Chapter 31 of the Book of Proverbs: "...She doeth him good and not evil all the days of her life."

INTRODUCTION

Like my previous book, **The Machinist's Bedside Reader**, this is a book for machinists and gunsmiths, both amateur and professional. All will find something of interest and value herein.

It is assumed the reader has a copy of **TMBR**. (If you don't, you will probably want to get a copy; references are made in this book to various ideas in **TMBR**. See inside back cover for details on how to order your copy.)

Comments made in the Introduction to **TMBR** still apply: I still have many things to learn, and you may not agree with all that I say.

Other than that, little else need be said except "Thank you" to all those who purchased **TMBR**, and a double "Thank you" to those who wrote or called to say that they enjoyed it. That, and your continued support as evidenced by the fact that you have come back to buy this new book, is much appreciated, and means a great deal to me in terms of personal satisfaction.

Although **TMBR#2** is not "the same" as **TMBR**, I hope that you find it as good or better, for that is what I have tried very hard to make it.

Some terminology: The symbol ℄ means "centerline"; Ø means "diameter". CRS = cold rolled steel; HRMS = hot rolled mild steel. Drwg = drawing; Drwn = drawn. OAL = overall length. shcs = soc. cap screw = socket head cap screw. hsm = "a home shop machinist".
HSM = Home Shop Machinist Magazine. M.E. = Model Engineer Magazine. GBL = me - the guy what writes the book.

I don't see myself as "the author" (or as "an author" - or worse still, the/an Author), writing something to be handed off to a publisher and that's the end of it. I view each project (this book, or the drawings for **Lautard's OCTOPUS**, or whatever) more like a visit with all "my guys" - I imagine us sitting around the stove in our collective "shop" at coffee time, talking, and I'm saying "I think you guys might like this... and Oh! Hey - I gotta tell ya about this neat whachamacallit a fella showed me the other day.. and lookit the drawings one of the guys down in Arkansas sent us - we can make one of these for doin' (something).. What do you think of it? - I think it's neat." And so on. Maybe that's why I put things the way I do when I write.

NOTES RE DRAWINGS

Each drawing in this book, was, in the course of preparation, judged by one standard only: "Could I make this part from what the drawing now shows?" If not, I put some more info on it.

However, it is important that the book text be regarded as an integral part of the drawings. Don't assume that just looking at the drawings, without reading the accompanying text, will give you all the info you need - *it won't*.

My drawings do not reflect certain drafting conventions, although I realize such exist. It is important for you the reader to know this, and I will explain why:

In industry, normal drafting practice is that if a dimension is given in inches and fractions of an inch, then the tolerance on that dimension is, unless otherwise stated,

±1/32". If the dimension is given as decimal inches to one decimal place, e.g. 5.6", then the tolerance is ±0.01"; if given as decimal inches to two decimal places, e.g. 5.62", then the tolerance is understood to be ±0.005"; if given to three decimal places, e.g. 5.625", then the tolerance is ±0.001".... Or something like that.

I do not use this convention. What I do do is expect the reader to have a common-sense understanding of what sort of tolerances would be appropriate for a particular part.

Say you are making the small Pantograph Engraving Machine. The Spindle Arm, Part #9 (page 31), is shown as having an overall length of 11-3/8". Now obviously, ±1/32nd is fine on the OAL. But on the same drawing, the spacing between the several pivot holes is also given as whole inches and fractions of an inch. These holes must be located just as close to the given nominal dimension as possible; it should be obvious, from the briefest consideration of the part these holes play in the finished machine, that a high degree of accuracy in locating them is required. Therefore, here 1-1/2" means 1.500", ±nothing if you can do it.

I often deliberately omit dimensions that are easily obtained by adding or subtracting others that are given. On the other hand, dimensions are not omitted if the reader would have to resort to higher mathematics to obtain them.

If a part appears to be symmetrical about a centerline, then it is, unless otherwise noted. Example: if two holes are shown as 4" apart, and appear symmetrically placed with respect to the part's centerline, then they are each offset 2" from the centerline, even though that info is not given. If it were not so, it would be dimensioned accordingly. All dimensions are in inches.

Radius-ing or beveling of corners is not called out - it is assumed that you will do this as a matter of course. Similarly, I rarely draw in what any machinist would know must be incorporated - a prime example of this would be a relief groove or undercut at the bottom of a hole to take a bearing, for example the ball bearing jockey pulleys for the lathe overhead drive, at page 44. These things are expected to be in the reader's repertoire already.

My Shop: It might be useful for you to know the type and size of the equipment in my shop, as this colors my thinking in describing how to do a job. My lathe is a Myford 'Super 7', which has a 7" swing (10" in the gap), takes 19" between centers, and will not quite pass a 5/8"Ø bar through the spindle hole. My mill is a knee-type vertical, with a 6x24" table, Taiwanese made, but quite nice. There's a 3/8" capacity bench drill press, also Taiwanese, which is rarely used. The rest is hand tools.

Wherever you are, and whatever you do in life and in your workshop - power to your elbow, and may your bearings never seize. I hope you enjoy **The Machinist's Second Bedside Reader**.

August 20, 1988

Guy Lautard
West Vancouver, B.C.

GUY'S COMPREHENSIVE INDEX TO THE BEDSIDE READERS

The best way to find what you're looking for in **The Machinist's Bedside Readers** is to look it up in "Guy's Index". 10+ pages. Covers all three Bedside Readers and "Hey Tim...". Price: US$7. Within Canada, Cdn$9.95, GST included.   Re "Hey Tim..." – see page 206

And write for our full catalog giving details and pricing of this and other books and videos, **some of which** are described at pages 203-207 of this book.

Some Comments on Originating Reference Tools

The machinist requires higher standards of accuracy in respect to certain basic geometric elements than most other tradesmen and artisans. He needs right angles that are 90°, no more, no less. He needs - though not always - surfaces that are flat to within a tenth the thickness of a cigarette paper. He needs to be able to divide a circle into any number of precisely equal spaces he may require. He needs to have at hand, and/or be able to produce, items of almost perfect roundness, threads of any pitch required, and so on. What would have delighted a carpenter, or a wagon maker, or a windmill maker (millwright) was worthless to even the early day machinist, and it is only more so today.

Today, one can buy an extremely accurate square as readily as a newspaper. A dividing head can be bought just as readily, complete with a set of division plates whose accuracy need never be doubted; the price may be about the same as a set of tires for a car - or less, or much more.

My point is that such things are readily had today. But what about 250 years ago, when the Industrial Revolution began? Where then did you buy a Laboratory Master granite surface plate? Men like us, but far less well equipped, in terms of shop tools, mikes, lathes, etc., had to figure out ways to originate such things from first principles, and to sufficiently high standards to serve the needs of a new and demanding field of industry. Then they had to set about to carry those new ideas into reality.

In the pages that follow, we will look not at "how it was done", but at how we can do some of these things ourselves, should we have the need and/or the inclination to do so. There is no point spending days making a surface plate when one can be bought for 3 or 4 hours wages. But some of the things we will look at make good sense to know how to do.

And if not all of these are of interest to you, you may be sure someone else is probably delighted to find such material herein. So don't begrudge the space they take up... if not of interest, turn the page - there, I hope, you will find something of value to you.

A SUPER SIMPLE INDEXING DEVICE

The indexing device shown at Fig. 1 below is simplicity itself. I saw this idea in the October 1984 issue of The British Horological Journal, in a clock-making serial written by Eliot Isaacs. I contacted BHJ and the author, and obtained their ok to work up my own version of it for use herein. I plan to make one like it, but slightly heavier, for my own shop, and I expect to get good use out of it when I do. I intend to make a bigger one as well.

As presented by Mr. Isaacs, the device was designed to attach via two T-bolts to the vertical slide of a Myford lathe. It could also be bolted to an angle plate for use on the vertical mill. Alternatively, it could be made to be held in the milling machine vise, as at Fig. 2, or mounted to a lathe milling attachment. In any case, I think it is an excellent idea which can be adapted to many different readers' activities.

It is likely that most would want to beef up the design somewhat from the dimensions given here. Although Mr. Isaacs is a highly qualified clockmaker, and has used this device for making some of the parts for several clocks, it strikes me as being somewhat lightly constructed throughout; for example, Mr. Isaacs bolted the two main pieces together with two 2BA (same as 10-32) socket head cap screws. Just how heavily built such a device should be also depends on the type/size of work to be done with it. Mr. Isaacs doubtless found his satisfactory for his purposes. I will probably use say 1/2" hot rolled mild steel plate for the vertical plate, 1/2" or 5/8" for the horizontal piece, and two 1/4-20 socket cap screws to join them. I'll use 10-32 (not 6-40) screws to clamp the arbor. I may get it built and decide it is too heavy all around, but I doubt it.

Fig. 1 A Simple Light Duty Indexing Device

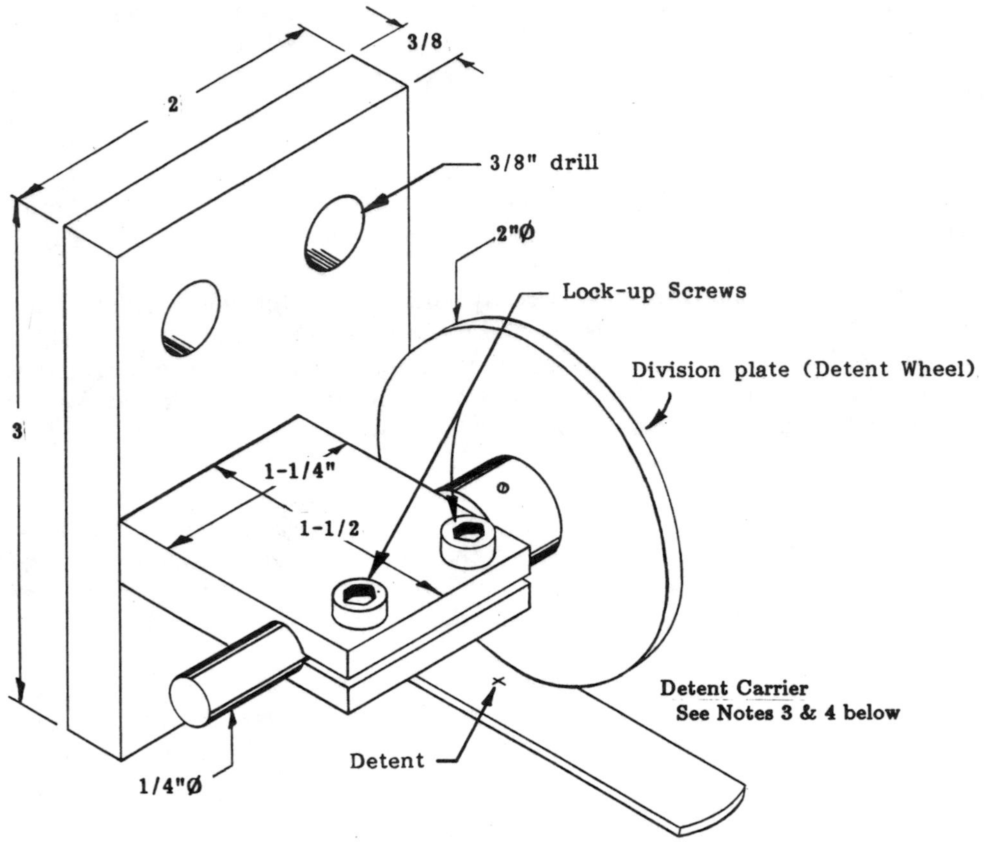

NOTES

1.   The two main parts of this very simple but entirely functional indexing device are bolted together with a pair of screws entering from the back of the vertical piece. The completed device is bolted to the lathe's vertical slide (milling attachment) with T-bolts through the two holes in the upper portion of the vertical piece.

2. The 1/4"∅ shaft can be the workpiece, as when making a clock pinion from 1/4" drill rod, or it can serve as the arbor for a workpiece mounted on it.

3. The detent carrier is a piece of 1/2" x 2-3/4" x 1/32" or similar springy steel (e.g. a piece of hacksaw blade).

4. NOT SHOWN are a pointed screw detent, and detent notches in the periphery of the 2"∅ detent wheel or division plate (too hard to draw!).

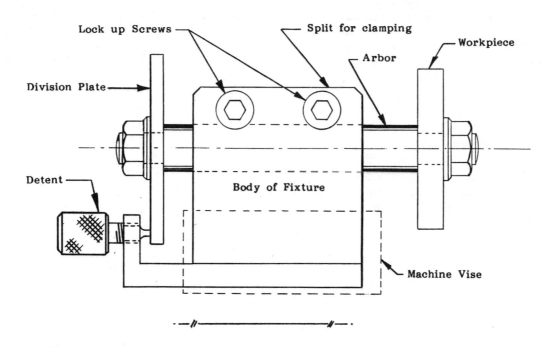

Fig. 2: A Beefed-up Version Mounted in a Milling Machine Vise

Vertical mill provides powered spindle for cutter

Lock up Screws — Split for clamping — Workpiece

Division Plate — Arbor

Detent — Body of Fixture — Machine Vise

Custom division plates

See **TMBR** for info on making division plates from bandsaw blades.* From such a starting point, it would be easy to make custom division plates for the Indexing Fixture shown above. Such plates could be a little more elegant than a bandsaw division plate, and sized to suit your particular rendering of the basic device. There is also a way of originating a master index plate from nothing, which we will look at next.

*see also a comment re same in Appendix A herein, for a footnote to one aspect of that idea. GBL

ORIGINATING A MASTER INDEX PLATE

The idea which follows is probably *the* classic method of originating an accurate division plate from scratch. Although not originated by him, Martin Cleeve described it very nicely, from his own practical use of the method, in an article he wrote in *Model Engineer Magazine* in the mid 1950's. The presentation below is based on that article, but I have added a number of points of my own. (Note: I have used the terms "Index Plate" and "Division Plate" interchangeably here. GBL)

If you need to make a very precise master index plate, you can do so quite readily with little more than your lathe. If you need a high number plate, there's going to be a fair amount of time involved, but that probably won't stop you if you want/need the result badly enough.

What to do? First, a bunch of arithmetic. Then make as many small buttons of a convenient size as you need divisions. A practical size to make the Buttons is something just under 3/8"Ø as they can then be conveniently turned out of 3/8" CRS. Make a backing or carrier plate, and a locating disc. The Buttons and the Locating Disc are sized such that when the Buttons are placed on the carrier plate around the periphery of the Locating Disc, they will all touch it and each other. See Figs. 1 & 2. (next page)

The Locating Disc is secured concentrically to the Backing Plate, and a tapping size hole is jig drilled and tapped through a Button into the Backing Plate. Screw the first Button in place. Jig drill and tap the hole for the second Button, and put Button #2 in place. Repeat till all the Buttons are in place.

Mount your division plate on whatever you want to use it with, and run an indexing finger in between adjacent Buttons, as at Fig. 3. Now you can make a working plate from your master.

Fig. 1

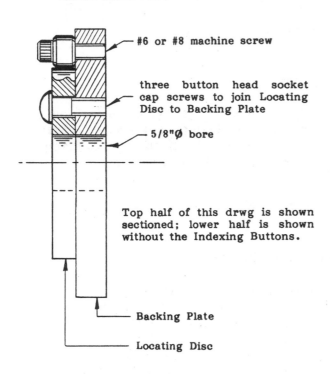

- Backing Plate Ø = B
- Button Ø = 0.365"Ø
- Socket cap screw
- Pitch Circle
- Locating Disc Ø = D

Assembly drwg of a Master Index Plate   **Fig. 2**

- #6 or #8 machine screw
- three button head socket cap screws to join Locating Disc to Backing Plate
- 5/8"Ø bore

Top half of this drwg is shown sectioned; lower half is shown without the Indexing Buttons.

- Backing Plate
- Locating Disc

Fig. 3 The Master Index Plate in use

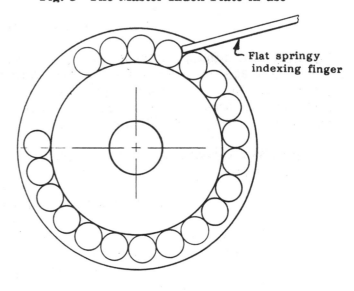

- Flat springy indexing finger

The arithmetic behind the method:

Let's take a look at an example using 1/2"Ø Buttons. (See Fig. 4, next page.)

Say N = Number of divisions wanted = 38
Say Buttons are 0.500"Ø
What should be Ø of Locating Disc?

$360°/38 = \alpha = 9.4737°$
$\Theta = \alpha/2 = 4.736842°$
$R = r/\sin \Theta = 0.25/\sin 4.736842° = 0.25/0.082579 = 3.027391"$

$R_1 = R-r = R - 1/4" = 3.027391 - 0.25 = 2.777391" = R_1$

Therefore, Locating Disc Ø = 2 x 2.777391 = 5.554782"

4

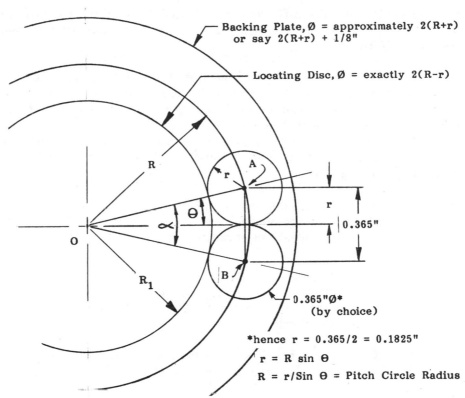

Figure 4

The Geometry which
allows us to
originate
a Master Index Plate
from elements
easily made on a lathe

Backing Plate, Ø = approximately 2(R+r)
or say 2(R+r) + 1/8"

Locating Disc, Ø = exactly 2(R-r)

0.365"

0.365"Ø*
(by choice)

*hence r = 0.365/2 = 0.1825"

r = R sin Θ

R = r/Sin Θ = Pitch Circle Radius

To find Θ, let the number of divisions wanted be N

Then ∝ = 360/N, and Θ = ∝/2, or

to get Θ directly, Θ = (360/2)/N = 180/N = Θ

Thus, for N = any number of divisions we may want, we can have
Θ and R in very short order. And if, for practical reasons, we
standardize on a button diameter of 0.365" for most purposes, we
can go pretty fast on the making of any master division plate we
may want, as well as on the aritthmetic.

Why not work up the numbers for a 27 division plate, and see if
you get the same figures as given below without the calculations?

N = 27   r = 0.1825"   Θ = 6.666,667°   R = 1.572,02"

Locating Disc = 2.7790"Ø, and Backing Disc about 3.6"Ø

Possible Errors:
Now suppose that the actual size of the Locating Disc produced is Ø = 5.555", which would be
0.000,217" oversize.

Ø/2 = R_1 = 5.555/2 = 2.777,500"

R_1 + 1/4 = 3.027,500" = R

r/R = 0.25/3.027,500 = 0.082576 = sin Θ, which we will call $Θ_1$.

$Θ_1$ = $\sin^{-1}$(0.082576) = 4.736672°

But we have already found that if all was dead on, $Θ_0$ = 4.736842°.

Therefore the error per space = $Θ_1$ - $Θ_0$ = 0.000,17°

How serious is this? Let's look into it:

If Ø = 5.555"Ø, i.e. 0.000,217" oversize, then R_1 + 1/4 = 3.027,500" = R

and the Buttons would need to have a diameter of

2r = R(sin Θ) = r = 0.250,009", which we cannot produce.

5

Can we neglect it? The difference is 0.250,009" - 0.25 = 0.000,009" per Button. This, multiplied by 38 Buttons, is 0.000,3", which we know we can quite happily neglect, maybe even if doing stuff for NASA.

I said above that a good size for the Buttons would be just under 3/8"Ø. One need not stick with a Button dia. of 0.365" although this is a good practical size to make from 3/8" CRS. Low number division plates can be made using smaller Buttons, but there are benefits to making a master index plate which is a multiple of the desired number: the Buttons remain a convenient size, and the final index plate may have greater versatility because it has more spaces, while still incorporating the number you wanted. (If you use a different Button size, change "r" accordingly in Fig. 4.)

It is preferable to keep the master division plate's pitch circle close to, or larger than, the diameter of the final workpiece to be divided out from it. If this is done, errors will be reduced, whereas they would be magnified if the master were say 1/2 the size of the job.

To make the Buttons (See Fig. 5.)
Face the end of a piece of 3/8" CRS. Break the sharp outer edge with a file, and part off a piece a little longer than finished length, say 0.257", for a final length of 1/4". Repeat until you have more Buttons than you need.

Chuck each in turn in the 3-jaw and center drill, drill, and ream 3/16". (This assumes your 3-jaw will center a piece of round stock within say 2 or 3 thou; if not, adjust your methods accordingly.)

Next, chuck a piece of 1/2" CRS, with about 5/8" protruding. Face the end, and turn down for about 3/8" to a good twisting hand force fit in the reamed hole in the Buttons. Undercut the root as noted on Fig. 5, and face out to leave a nice square shoulder for the Buttons to butt up against.

Push on a Button, parted end out, face, and turn to specified dia. with a tolerance of plus 0.000,2", minus 0, and put a slight chamfer on outer corner. Do all the Buttons thus.

Leave the chucking stub in the lathe, in case you have to polish a hair off the Buttons.

Figure 5: Sequence of machining operations for making the Buttons

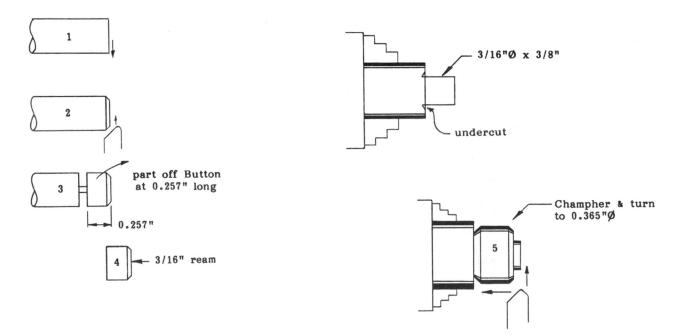

6

Backing Plate and Locating Disc (See Figs. 2 & 3 again)

The Backing Plate and Locating Disc (which should be made up before the Button chucking stub is made and used, because, as noted in the preceding paragraph, we want to be able to leave the chucking stub undisturbed in the chuck for a while) can be made from 1/4" CRS plate, or HR plate faced both sides.

Locate center of blank, chuck, indicate for zero face runout, and with center running true, drill and ream say 5/8" dia. Drill and tap 3 holes in Backing Plate at some convenient radius, and three matching clearance holes in Locating Disc. Assemble with button head cap screws, or similar. (Alternatively, solder or rivet the two parts together, and then ream the 5/8" dia. hole.)

Assembly:

Put a Button in place against the Locating Disc, and drill a hole (in the Backing Plate) for a #6 or 8 machine screw. Tap this hole, and screw the first Button home. Put next Button beside and repeat. When all holes are drilled and tapped, and all Buttons assembled, you may find the last one won't quite go into place. If so, loosen several adjacent Buttons, and pull each one into tighter contact as you re-tighten them. You may have to do this at several places around the circle before the job is done.

In his description of this method of making a Master Division Plate, in *M.E.*, Martin Cleeve suggested making a final check on one's work by turning a chucking stub to fit the 5/8" dia. hole, and mounting the Master Division Plate thereon in the lathe; having done so, he suggested taking an indicator reading on the periphery of the Buttons - the max. reading should be the same for each.

Finally, copy the Master, either as a conventional division plate with holes, or as a ratchet type, or whatever other form suits your purposes. A finger of flat steel - possibly a piece of hacksaw blade - rigged as shown in Fig. 3 will work well as an indexing finger between the Buttons on the Master Division Plate.

Best way for our purposes, if doing a drilled division plate, would be to drill each hole in the copy plate with a good sharp center drill, going in deep enough to form a chamfer for the finished hole. Once all the holes are thus spotted, dismount all, put down a piece of masonite board to protect the drill press or milling machine table, and finish drill the holes with a drill of the appropriate size for the detent plunger - 3/32" dia. is a good size for the latter.

VERNIER DIVISION

There is a way to produce a desired number of divisions from a simple indexing device when a plate containing that number is not available. A second plate is introduced, and each plate carries a series of divisions which together will produce the desired result.

In *M.E.* for Nov. 6, 1970, "Mac" Mackintosh described this method. Mac had made a simple indexing milling fixture to drawings published in *M.E.* for Oct 31, 1968 by G. Douglas, and in his article told how he had modified the Douglas design to utilize this principle.

Say we want 45 divisions. We make one plate with 9 holes on a pitch circle of some convenient diameter. (Or 9 notches in the rim.) We make another plate with 5 holes on a pitch circle of some other diameter, and from this we jig drill/ream one additional hole in the 9-hole plate, so a pin can be put through it into any of the holes in the 5-hole plate - obviously the 2 plates are made to mount concentrically on the dividing spindle arbor (= work spindle). The holes might all be reamed say 3/16" dia., for use with a "dowel pin" made from 3/16" drill rod.

The low number plate is locked to the work spindle, and the high number plate is free to rotate on the same axis. See the drawing below.

A start is made by putting a pin through the single hole in the 9-hole plate and into the #1 hole on the 5-hole plate. Both plates are thus locked together, and, via the low number plate, to the work spindle. We then proceed to cut 9 divisions. We then remove the pin, and rotate the 9-hole plate so that the pin can be put through it into the #2 hole in the 5-hole plate. Cut 9 more divisions, and repeat. Pretty soon, 45 divisions.

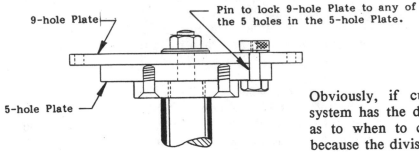

Obviously, if cutting graduations on a dial, the system has the disadvantage that no clue is offered as to when to cut longs and when to cut shorts, because the divisions are not cut in sequence.

This vernier principle can be applied in other numbers than 45. Some examples:
- for 60 divisions: 12 and 5.
- for 100 divisions: 25 and 4.
  (for 50 divisions, same, but use 2
  of the 4 holes on the vernier plate)
- for 288 divisions: 32 and 9
- for 360 divisions: 40 and 9

The number of holes in the two related plates must not be divisible by the same number. If a prime number is desired, the vernier system will not work - you must make a plate having the required number of divisions.

NATURE OF THE DIVIDING DETENT

The nature of the detent finger can vary depending on what you are doing. A ratchet form of division plate is fast to operate - consider the bandsaw blade division plate idea in **TMBR**, page 43. Holes in the division plate can be radial or axial, and if radial, notches can be used instead, where they will be faster to use than a peg-in-hole arrangement. If a peg is used, it can be tapered, which helps both speed and accuracy; it can also be spring loaded. The end of the peg should be well rounded and polished; if this is not done it will scratch the face of the plate as it is moved from hole to hole.

A QUICK SETTING 1/4° VERNIER

Assume we have engraved a 0-360° protractor on something, and we want a vernier that is fast to read and set, and which will provide us with settings to an accuracy of say 1/4°; for many applications such a level of accuracy is entirely adequate.

If you make an index plate with an inner radius matching the graduated O.D. of the protractor, and cut thereon a vernier scale consisting of 5 graduations spaced 4° 45' (i.e. 4.75°) apart, and number them 0, 1/4, 1/2, 3/4, and 1, you will have a vernier which reads directly and quickly to 1/4 degree. Not as precise as the 5' vernier, but probably 3 times as fast to set, and 4 times as easy to see which mark is getting lined up.

Think about it for a minute: a 5' vernier spans about 11° and has 12 marks in it. A 1/4° vernier has only 5 marks in it, and spans 19°.

Here's a good idea for
ROUGH DIVIDING

Engrave a series of axial lines, say 5 or 6 thou deep, at intervals (see drawing at bottom of this page) on the rim of your chuck backplate, faceplate, or whatever. Number them 0, 30, 60, 90, etc. and dress down the burr thrown up by the stamps. Make up a little chart as follows, and put it up near your lathe.

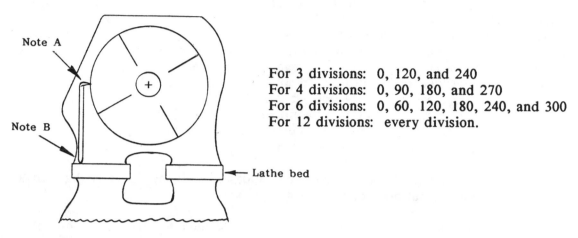

For 3 divisions: 0, 120, and 240
For 4 divisions: 0, 90, 180, and 270
For 6 divisions: 0, 60, 120, 180, 240, and 300
For 12 divisions: every division.

Note A: Sharp point to enter lines engraved on rim of chuck backplate. Length is not critical.

Note B: Opposite end can be broadly radiused, and polished, to prevent possible damage to the lathe bed.

To use, align each of the appropriate graduations with the point of a bent scriber as in the drawing above.

While obviously not for high class dividing, this method will do quite nicely for some jobs. I intend to graduate the rim of every faceplate and chuck in my shop; by doing them all at one session, while I have the geared rotary table set up, the amount of time required to do each one will not be excessive.

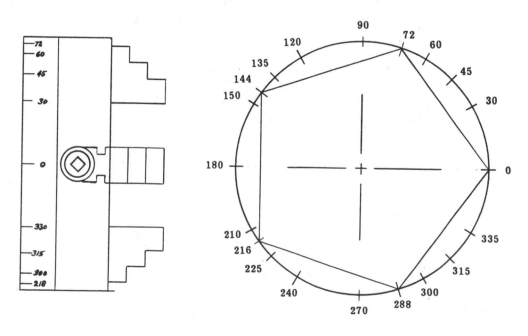

9

HOW TO MAKE
A MASTER REFERENCE SQUARE
(Adapted, with permission, from an article
in the July 31, 1924 issue of MODEL ENGINEER Magazine.)

(I found this article in the July 31, 1924 issue of MODEL ENGINEER Magazine. The author's name is not given. He wrote in a manner and style that would be difficult for the average American reader to follow. I have therefore edited quite freely throughout, although where it would provide additional interest I have left some of his terminology intact. If I might liken the writer's presentation to that of a butler serving dinner, I would say that he seems to have entered the dining room walking backwards, but it turns out to be not a bad approach after all, so I have left it in much the same order. However, his last paragraph I have moved to the very beginning: it is a warning to some that they would be wasting their time reading this item, for not all will have the inclination to undertake such a project. In spite of that, it is included here because I am sure it will interest some, and because it describes methods and reasoning useful to anyone who may have occasion to true up an out-of-truth machinist's square, or to create - from nothing, as it were - an accurate square, a straightedge, a surface plate, a division plate, or like that. If interested, read on..)

(There follows immediately after this item another, showing a somewhat different approach to a similar, but not exactly identical problem.)

NOTE: Both methods would also be useful if one wanted to true up an out-of-truth machinist's square.

HOW TO MAKE
A MASTER REFERENCE SQUARE

Only a genuine tool-lover will take the pains necessary to make for himself a master reference square. However, this can be accomplished, using only a true straight edge as a reference. In the process of making a Master Square, three squares are in fact originated. The straightness of the edges produced in the various squares is critical, for it is by this factor that the error of squareness, if any, is detected and corrected.

What is a Master Square?
A Master Square is one made to the highest possible standard of accuracy, and reserved exclusively for testing other squares, and as a first reference for testing, setting or truing other tools or gauges, the latter then being used for routine testing of work in hand. Consequently, a Master Square should be made in the simplest manner possible, and, in order to avoid any deformation due to hardening or case-hardening, it should be left soft.

Being soft, such a square must be provided with an extremely well made protective case, to guard it not only against knocks and dings, but even from damp or other effects that might lead to rusting of its finished edges. The case should be made of wood, and so designed that even if the case is dropped, the square will not be damaged.

It goes without saying that a Master Square should not be subjected to any kind of rough usage. The greater the degree of accuracy to which it is made, the more likely it is to suffer deformity from shock or even vibration.

Simple Form of Master Square
A Master Square for general use should be made not unlike a small carpenter's square, from a single thickness of flat steel sheet - see Fig. 1.

It should be made just slightly larger than the largest square to be checked against it. Beyond this, the smaller it is the better, because the smaller it is the less liable it will be to be knocked out of square by accident.

A good size for a Master Square might be 7" x 4" on the inside edges. But if a smaller one will serve you, then make it smaller, as noted above. At 7" x 4" size, the width of the blades should be about 7/8" or 1", and the thickness not less than 3/32".

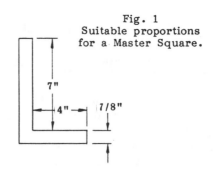

Fig. 1
Suitable proportions
for a Master Square.

Charcoal iron (see next paragraph) is a good material to use, because it is homogeneous and very easily worked, and can be obtained in nice flat sheets, with a fine black coating of oxide on each side. This oxide forms an excellent protective surface finish for the flat sides of the square, and should not be removed, thus also saving much work in finishing the square.

> (I phoned my favorite steel supplier and asked what is charcoal iron? Naturally, none of the young bucks knew, so they put me onto an old gent they keep stored away in a little room upstairs. I read him pertinent bits of the foregoing, and he told me the closest thing to charcoal iron today would be ordinary mild steel sheet. That's good news, because mild steel sheet is relatively easily come by. Come to think of it, I have some pieces of mild steel sheet about 1/8" thick stashed away under a bush in my shop where nobody is likely to spot it and ask for it for casual purposes...GBL.)

> NOTE: steel suppliers generally refer to material under 3/16" thick as "sheet", and material over 3/16" thick as "plate".

No stamping of letters or figures should be done on the metal either before or after the square is made. Marking, if any, should be done only by careful etching, or by hand or machine engraving, preferably before the square is made.

To Make a Master Square
For the size suggested, obtain a piece of hot rolled mild steel sheet about 9-1/4" x 5-1/2", cut as square as possible.

> NOTE: The piece should not be sheared to the desired size, but rather it should be requested cut say 1/2" oversize both ways. If you can get it cut with a coldsaw, or machined to size, so much the better - this will save some work with a hacksaw. The average hsm is likely to get his material from a steel fabricating shop. The shop may have a suitable nibbler or circular saw with which to do this. However, don't push your luck: better a piece of material, and some hacksawing work to do at home than to ask for the extra favor and be told, "Get lost, we're too busy."

Fig. 2 shows how to mark out the sheet. The portion marked "A" will become the Master Square, and "B" will be one of the checking set-squares, as explained below.

In addition, you will need a piece of nice flat "sheet metal", say about No. 20 S.W.G. (= 0.036" thick) or heavier. This can be tinplate - i.e. tin plated sheet steel - from a sheet metal shop. For the reason given in the next paragraph, this piece should be a little bigger than the piece of mild steel sheet, and should be cut approximately square on one corner. From this will be made a second set-square "C", not illustrated, but similar in shape to that shown at "B" in Fig. 2.

It is necessary to preserve the flatness of the tinplate material. Therefore, when you go scrounging about for this piece, take a pair of tin snips with you, and snip the extreme corners off upon receipt so that they will not be so readily bent. (This is the reason promised above: it must be bigger to make up for the amount lost in snipping the corners off.)

Set-square "C" should be made about the same length as "B" on the long side. The angle of slope of the set-squares does not matter, nor is it necessary to finish the sloping side perfectly straight and true - this is a non-working edge, "out in the wind", so to speak.

If you find you can cut the sloping side of set square "C" on a shear without putting "out of flat" the piece you need, then by all means shear it, but it is unlikely you will find this possible. In any case the material in "C" must, in the final analysis, be flat.

There is nothing to say that "C" must be triangular in shape. It could instead be made as a rectangular plate. However, in this form it will be found less convenient, because the corner which you will make truly square is not so easily identified, there being then four of them instead of one.

Reference Tools Required
As stated before, this is to be an original square, (i.e. one originated "from scratch") and therefore no reference squares other than those produced are needed.

What you will need, however, is a good straight edge, say 1-1/4" x 1/4" and 12" or more in length. For convenience, an accurate surface plate will also come in useful, say about 12" x 9".

The straight edge can be a draftsman's steel straight edge, or something else that will serve the purpose, such as a length of aluminum extrusion, tested for straightness, and corrected if necessary. (By the time you have read this through once, you will not need to be told how to true up the edge of a straight edge. GBL)

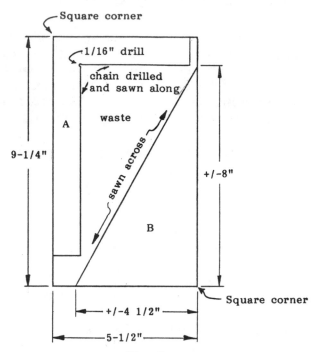

Fig. 2
Layout for cutting the Master Square and its checking Set Square from the original piece of 1/8" hot rolled mild steel plate, after squaring up two corners as at Fig. 4 below.

An excellent surface plate for the purpose is a piece of plate glass of about the size noted above, preferably with its edges ground square and true. A piece 3/4" thick is fine. If you can get one thicker than this, so much the better.

A glass surface plate should be mounted on a piece of flat plywood or heavy masonite board. The board should be quite thick - say as thick as the glass - and fitted with three feet underneath so that it will sit steady wherever it is put down, on the workbench or elsewhere. The board should also have edging pieces on all four sides, so dimensioned that the top surface of the glass stands above the edging.

When so mounted it is tried against the straight edge all ways, and if any dishing of the plate is detected, one or more pieces of thin paper put between the glass and the wood at the center will generally correct the tendency to sag. If on the other hand it is found to be high in the center, the error can be corrected by some judicious work with a plane (or sandpaper) at the center.

> (My own surface plate has been, for about the past ten years, a 1 foot square piece of 1" thick plate glass. In use or idle, it sits on a scrap of carpet, the latter being placed directly on the workbench when I am using the surface plate. My theory is that the carpeting will serve to iron out any possible unevenness of the workbench top which might otherwise tend to distort the surface plate. More recently, I have acquired a Starrett "Crystal Pink" Toolmaker's Flat. It is about 12" x 8", and 2" thick, so it is not going to sag. See further notes on the subject of surface plates elsewhere herein, and see page 8 in **TMBR**. GBL)

The General Procedure
The first step is to rough out the Master Square and create one finished edge in the manner to be described below. A start is also made on the two reference, or checking, set-squares.

The Master Square is then placed with its short edge on the edge of the straight edge, or on the surface plate, and is corrected on the long edge to agree with "B", the thick set-square, long edge to long edge, the latter also in contact with the surface plate or straight edge, as shown in Fig. 3, at right.

When this agreement is satisfactory, take up "C", the thin set square, and, in the same manner, make its long edge agree with the long edge of the Master Square. Naturally, the Master Square is not altered in this phase of the proceedings, but is used as the reference item.

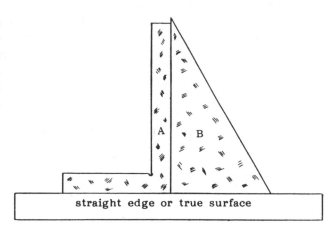

straight edge or true surface

Fig. 3
Trying the Master Square
against the first Set Square.

The two set-squares "B" & "C" are next set facing each other, and if they show agreement - a state of affairs which is highly unlikely at this stage of the game - the Master Square is as correct as it is possible to make it.

If, however, the two checking squares B & C do not agree with one another, the error shown is double that of the Master Square, whose error is the inverse of the error between B & C. For instance, if B & C touch at the top, and show light at the bottom, then the Master Square is high at the bottom, by half the error. If the set-squares touch in the center, the Master Square is hollow in the center by half the error, and so on.

The master is then "corrected", using judgment only. That is to say, it is corrected about as much as you figure necessary, based on gut feel, rather than by comparing it with something else that might be square. The corrected edge is next again made straight, by the methods to be given below.

Having done the two steps given in the preceding paragraph - i.e. correcting and straightening - we must once again go through the procedure of making B agree with A and then C with A and then testing C against B, in that order.

This process is repeated until all three squares agree perfectly with one another, whereupon the job is finished so far as the outside edges of the Master Square are concerned.

To finish the inside edges of the Master Square, the shorter side is first finished straight and parallel to the outside by micrometer (or with a sensitive dial indicator) and the long side is then finished straight and true to micrometer and checked for squareness against the two checking set-squares. Because the two checking squares are known to be square - we have made them so - any lack of agreement between them and the inside of the Master Square is corrected by making modifications to the Master Square only, until finally the Master Square is truly square inside and out.

After this happy state of affairs is secured, the thick set-square is cased with the Master Square as a permanent test piece, and the tinplate square can be scrapped, or its sloping edge trued up, and promoted to use in the drafting department.

The foregoing is the procedure we will use. The following paragraphs tell how to actually work on the various edges of the three squares.

Dealing with the Plate in the First Place
The first thing to do with the 9" x 5" mild steel plate is to true the two ends. Make these

13

straight by trial and error against the edge of the straight edge, using DyKem Hi-Spot Blue or equal, and working in the same manner as when scraping one surface true to another surface which is known to be true.

Lest there be someone who does not know about scraping one surface true to another, herewith a four sentence rundown:

A thin coating of gooey, slow drying Hi-Spot Blue is smeared on the reference surface, the surface to be made true is placed on the blued surface, and the two are rubbed together and then separated again. The blue will be transferred to high areas on the workpiece. These must then be scraped down. This is done a little at a time, with the workpiece being wiped clean and returned to the blued reference surface frequently during the process.

For more info on the art of scraping, see the very fine book, *Machine Tool Reconditioning*, commented upon at page 67 herein, in the section on Spot Grinding and Lapping. GBL

However, we are truing an edge, not a surface, and the work is not done by scraping, but by careful draw-filing, using smooth and finally dead smooth files. A dead smooth clock pivot burnishing file would be about the best for finishing. (This and the next-named item can be obtained from watchmakers' supply houses - see page 122. GBL)

Finishing can also be satisfactorily carried out using emery sticks, which are strips of blue-black emery paper glued to planed hardwood sticks. (See also the item re 1200 grit wet & dry paper, elsewhere in this book. GBL) Do not endeavor to finish with emery paper or cloth wrapped round a stick; this will result in rounded edges, which will be useless.

NOTE: See the idea for a cylindrical cast iron lap in the article which follows this one. The use of such a lap should make it difficult to produce anything except square edges. GBL

When true and free of dirt, spotting blue, etc., an edge should show an even line of light when viewed standing on the straight edge.

You will recall that we first set out to make two true edges, one at each end of the plate. Next square up and true the adjacent side edges of the plate. The truing up is done as for the end edges, but we must now begin to check also for square, as follows:

Having decided which side is to be made square to which end, clamp the straight edge on a flat board, or drawing board, with a sheet of white paper underneath, and test for squareness of two adjacent sides as shown in Fig. 4. Put the short end of the plate against the straight edge. Using a sharply pointed hard lead pencil against the long edge, which must be as nearly straight as can be got without finishing, draw a fine pencil line on the paper. Then turn the plate over, and with the long side near the pencil line, draw another line about 1/32" or so away.

Fig. 4
Testing a corner of the plate
for approximate squareness.

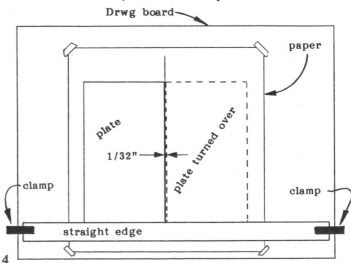

14

This is a severe first test for squareness; an error of less than a thousandth of an inch will be revealed if the long side is straight. When the lines show parallel by eye it will be near enough, but the pencil must be manipulated carefully, being held close to the edge with its top pointing away and kept at a constant relative angle.

True both sides of the plate square to their respective ends in this manner. Note however, that the intent is not to create a perfect rectangle, but rather only that two opposite corners be made right angles. The other two we do not much care about.

After this, mark off the plate in relation to the squared corners, as shown in Fig. 2, so that a corrected square corner forms the corner of the Master Square and of the set-square respectively.

Next, drill a hole about 1/16" or a little larger, exactly centered on the inner corner of what will become the Master Square. Thereafter the piece marked "B" may be hacksawn off, keeping roughly to the line, and afterwards filing the sawn edge reasonably true for good appearance.

To cut out the piece which will become the Master Square, drill, as close together as possible, a series of 3/16" holes beside, but just clearing, the layout lines. Finish this operation with a hacksaw, working as close to the layout lines as is practical.

To hold the work for some of these operations and for the subsequent truing up of the edges, a pair of flat-planed slabs of oak or similar hardwood would be useful. The plate is held between these, while the slabs are gripped - not too tightly - in the vice.

NOTE: To reduce the risk of dropping the square during this phase of the work, I would fit these two chunks of wood with spring loaded bolts and wing nuts. This would reduce from 6 to about 3 the number of hands needed to get the square and the two pieces of wood placed just so in the vice, and the latter tightened up. As you are going to be doing that little routine quite a few times, it is well worth the time this will take to arrange. GBL

In the paragraphs above headed "The General Procedure", most of the subsequent operations are described, and it only remains to point out that it is not much use to make any test on the surface plate unless it is known that the workpiece edges are themselves straight. This of course is tested against the straight edge.

The actual testing of the set-squares against the straight edge, and the testing of one set-square against the other, is done by observing the evenness of the line of light passing between the facing edges. This, if done against a good diffused light, will result in quite sufficient accuracy. Much depends upon the straightness of the edges, and if a good deal of trouble be taken to make the edges perfectly straight, the line of light showing between any given pair of edges can be reduced to nothing, but this depends upon the amount of nice work the maker is prepared to put into the job.

It goes without saying that the attainment of accuracy in this work depends to a great extent upon much wiping away of dust, and - as far as possible - working with clean hands, and the use of clean waste or similar material for wiping of surfaces and edges.

A Storage Case
The finished Square deserves a good storage box. This can be made from a slab of mahogany about 1" thick and large enough to take both the Master Square and the thick set-square lying in the case in about the same relationship as in the original layout of the plate. Batten the ends flush and make a lid, 1/2" thick and similarly battened. The two squares are let into the slab in separate, closely fitting depressions or beds. These depressions can be milled very neatly using a vertical mill, or can be made by careful drilling and chiseling.

Line the inside of the lid completely with a piece of baise or carriage cloth. Glue it on and allow the glue to dry thoroughly. Similarly, line the depressions cut for the two squares. Flush-hinge the lid to the slab. The tools are then laid in their depressions, on wool-packed cloth or velvet-covered pads.

3/8" holes can be drilled through both squares, making it convenient to lift them out of their beds.

(So says the original author. I would prefer to include, in the course of my chiselings, a finger relief beside each depression in the case. It would also be a good idea to include a small cavity for a piece of chamois cloth, the latter to be kept in the case with the Master Square, and used for handling and wiping same. GBL)

To Use the Master Square

To test any other square against the Master Square, two methods should be employed. Not only should the Master Square be applied inside and out to the item being tested, but both should be tested by standing them on a true edge or surface plate with their vertical sides matching.

The thick set-square B will be found useful to test the insides of small squares, but its chief use, and why it is retained, is that due to its superior stability it forms a permanent reference with which to check the Master Square if the latter is ever thought to have sustained an injury.

Neither square, however, should ever be left knocking about the bench or shop, and should never be used for regular work, which is better done by the use of hardened steel squares, which are usually corrected after hardening, by lapping them to a blackout fit against the Master Square.

MAKING AND CHECKING A FLAT SQUARE
By John J. McHenry
Reprinted, with permission, from
American Machinist Magazine, May 14, 1931, page 765

In making a device for generating parabolic curves of different focal lengths, it was necessary to make the flat Square shown in Fig. 1. Even in the best equipped shops, making such a Square is a fairly difficult job, and to our toolroom, making nothing but dies and special apparatus for automotive lighting, it proved interesting and in some respects rather novel. The essentials were a Square of truly 90°, the outer edges to be hard.

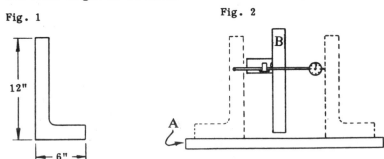

Fig. 1   Fig. 2

12"

6"

The Square was forged* from water-hardening steel, box annealed, machined to within 3/64 in. of dimensions, again box annealed, machined to within 1/64 in. of dimensions, and hardened.

*NOTE: I suspect that the use of the term "forged" here is completely inappropriate. I cannot imagine anyone setting out to hammer a piece of steel to make it flat and bring it nearly to the shape of a square, when his objective was to make an accurate final product. The Square is much more likely to have been cut from a piece of 1/8" water-hardening "gage plate", also known as "flat ground stock". If the writer's crew did otherwise, they made more work for themselves than they need have done. GBL

Before hardening, the Square was infolded in sheet metal 1/32" thick, leaving the outer edges exposed for a width of 3/16". Carbon steel will not harden in the section so covered, consequently when the Square was quenched the edges were glass-hard to a depth of 3/16", and the balance was soft. The warping due to hardening was easily straightened by peening, thus stretching the metal where required.

To aid in seasoning and removing any hardening and peening strains, the Square was plunged alternately in boiling water and ice water a dozen times, and allowed to remain in each bath long enough to reach the bath's temperature.

As to the grinding, I presume that the practice is much the same in any toolroom. We had the usual difficulty in keeping the Square flat, and finally corrected it by a very slight peening, unnoticeable but effective.

The method of checking and lapping may be of interest: Two parallels A and B were laid on a surface plate, as per Fig. 2. Parallel A was fixed, and Parallel B was also fixed (i.e. clamped immovable) during the testing operation. Placing the Square against Parallel A, alternately on one side of Parallel B, and then on the other, as shown by the dotted lines, and using a surface gauge having a sensitive dial indicator attached, the checking was proof of the correct position of Parallel B, was well as a check of 90° for the Square. Parallel B was used as a base to slide the surface gauge against, and so detect any errors in the set-up or in the Square.

For the sake of those who are sold on the use of a cylinder for checking a Square, I will say that it is not a satisfactory method for checking one of this size and thickness. The method of checking by parallels worked out nicely.

In lapping the Square, we fixed a piece of 1/2" x 2" cold-rolled steel across the face of an angle plate, as shown in Fig. 3. Clamping the Square against this piece brought it parallel with the face of the angle plate. The lap was a cylinder of cast iron, 1-1/8" in diameter and 2" long (faced off in a lathe, thus leaving the end square with the side of the cylinder. GBL). Using the end of the cylinder for lapping, and keeping its side against the angle plate, insured that the lapped edge of the Square would be at a right angle to its flat surface. The rest of the job was up to the workman's skill and ability, and in this case the result was highly satisfactory.

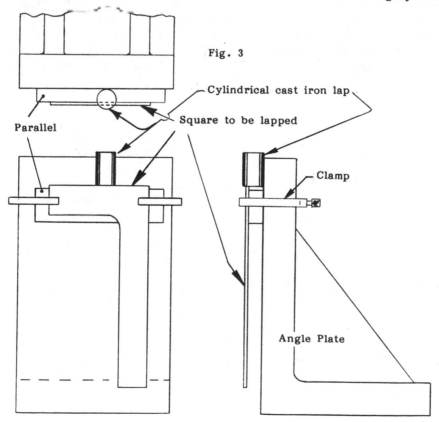

Fig. 3

Cylindrical cast iron lap

Square to be lapped

Parallel

Clamp

Angle Plate

17

A SHOP-MADE CYLINDRICAL SQUARE

Looking through Mac's letters to me, I came across the following item. It deals with the making of a cylindrical square, which the above writer dismisses almost out of hand. I would not presume to argue for or against either type, but will simply say that both probably have their place, or, to coin a phrase, "Different tools for different fools."

"At Sperry, all our reference squares were of the cylindrical variety. If you have *HSM* Magazine for May/June 1985, see page 32 and 33. There you will see a photo showing a square of this type, although no comment is made about it. Making one is as easy as falling off a log, and much more satisfying: one needs only to have suitable material - to wit, a piece of material of exactly uniform diameter throughout its length.

"Back in the 1950's I came across what I believe were surplus submarine periscope standards; they were 8 feet long, and consisted of a piece of thick-walled tubing with a sliding core to fit inside the tubing, all in stainless steel, beautifully ground, and in excellent condition. I bought two, at $5 each.

"One of the first things I made from this material was a cylindrical square. The outer tube was 3"OD, with a 2.5" bore. I sawed off a 9" length, being careful not to mar the ground finish, and mounted this in my very fine 6-jaw Griptru chuck with 8" protruding* from the chuck. I adjusted the job until the indicator showed zero runout close up to the chuck and at the outboard end.

"I then proceeded to face off the end with fine cross feed, light cuts and a very sharp tool, and then broke the inside and outside edges. I reversed the workpiece, and did the same for the other end. Placed on the surface plate and checked with an indicator, the ends were found to be strictly parallel to each other - my reference square was finished.

"By such means one can readily make a reference square which is superior to anything you can buy other than a professional cylindrical square at an enormous price."

> *That's a lotta overhang! I ran the first draft of the above past Mac, and he confirmed that it can be done, with fine cuts, a sharp tool, and much care. He also pointed out that if using a 4-jaw - which most of us would have to do, 6-jaw chucks being rare birds in basement workshops - it would be wise to make a plug for the inside of the tube, to prevent squeezing the tube out of round. Note too that if a cylindrical square less than 9" tall will serve your needs, make it shorter, thus cutting down on the overhang in the machining operation. GBL

A TRUE SQUARE

Comment: While you may not want to make a precision square of the type described here, or undertake any other work of this sort, which might properly be called gage making, you will nevertheless gain something - insights, ideas, and "grist for the mill" - just by exposing yourself to the techniques employed and shown herein and in the references cited in this and related parts of this book.

Starrett's Webber Gage Division makes high class gage block stuff, hence not cheap, and in fact for practical purposes beyond the reach of most hsm's. One item in this line is called a True Square. They make two sizes: the one the average hsm would find most useful, or get the most fun out of, is priced at about $800. However, you can make your own - you may consider it worth the time it'll take, given uses you may foresee for it.

A practical size of True Square for a hsm would be say 1/2" thick x 3" square, per drawing below. The one you or I would be most likely to want from Starrett's catalog is 5/8" thick x 4" sq., but I think that might prove to be a little big, hence the suggested 3" sq. size. (Their other True Square is 1" thick x 2" sq.)

We cannot hope to make ours to the same standards of accuracy as does Starrett, but if we can hold to ±0.000,1" on each side, accuracy will be ±7 seconds.*   (If we do no better than ± 0.001" in 3", and I think one could readily do much better, the angular error would be about 1 minute and 9 seconds.   Any patient worker with a small granite surface plate plus a "tenths" indicator should be able to work to much better than ±0.001", and if he does, he will have a True Square that is quite accurate enough for most of his purposes.)

*I don't want to bury you in mathematical garbage, but lest thee wonder "How did he figure that out?", here comes some:

Assume we make a True Square as at right, with an error on one side as shown:

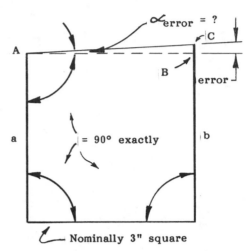

Consider the Triangle ABC:

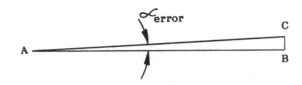

— Nominally 3" square

sides a & b are exactly parallel

We know that   AB = 3",   and that   BC = 0.000,1"

Then what is the angle α_{error}?

$$\alpha_{error} = \tan^{-1}(BC/AB) = \tan^{-1}(0.000,1"/3") = \tan^{-1}(0.000,033,333)$$

$$\alpha_{error} = 0.001,909,859°, \text{ which is about 7 seconds.}$$

A TOOLMAKER'S BLOCK

A Toolmaker's Block is a block of high accuracy for square, flat and parallel, to which a workpiece may be fastened. Operations then done on the workpiece while the Toolmaker's Block is placed in various orientations on (say) a milling machine or drill press table, surface plate, etc, will automatically come out similarly (if perhaps not *quite* equally) square.

This concept will need no further elaboration for any reader who has ever had a need for such a device. The drawings below give dimensions and salient features of mine. If your equipment is bigger than mine (see notes re same in the Introduction), and/or if your workpieces are expected to be larger, you will want to make your Toolmaker's Block larger - just make it good and square, for on this hangs all its usefulness.

NOTE: You may also be interested in **Lautard's Octopus**, a not-so-distant cousin to the Toolmaker's Block. The **Octopus** is a machined-all-over octagonal block of cast iron, part bench block, part indexing device, and much more. For details, see Appendix A.

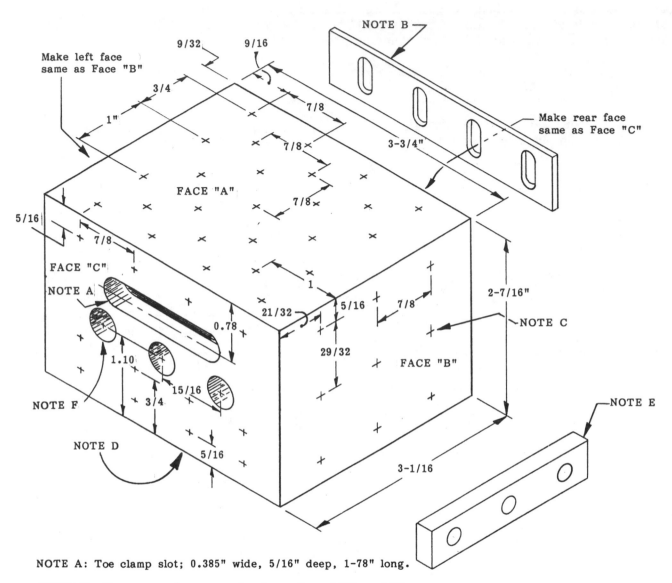

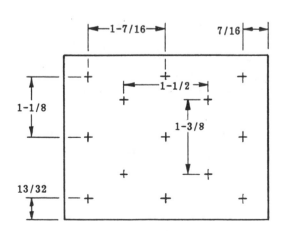

NOTE A: Toe clamp slot; 0.385" wide, 5/16" deep, 1-78" long.

NOTE B: Drop-away fences; make two, from 1/16 x 3/4" precision ground flat stock, one 3-9/16" long, with 4 slots, for long sides; one 2-3/4" long, for ends - this one will also be usable without end-overhang on long sides. Buy seven 10-24 x 5/16" socket cap screws to attach fences. Slots, pitched 7/8" apart, will match pitch of most holes in Block, including all edge holes. Make Slots 0.51" x 0.195".

NOTE C: Drill & tap 10-24 holes where this symbol appears, on 5 faces. Drill tapping holes 0.68" deep, except where otherwise noted. Assume symmetry where hole spacing is not called out. See Note D re 6th Face.

NOTE D: Re bottom face: drill 13 holes, all 0.8" deep, & tap 5/16"-NC, per layout at right.

NOTE E: Screw-attached clamping-down ledges. Make two, 2-1/2" long, from 1/4 x 1/2" CRS, c/w 3 holes drilled #9 (= 0.196") on 7/8" centers. Secure to Block, when needed, with three 10-24 x 3/4" socket cap screws.

NOTE F: On my own Toolmaker's Block, I drilled three 13/32" holes the full width of the Block, spaced so that the distance between the two furthest apart holes matches the 1-7/8" spacing of adjacent T-slots on my milling machine.

20

THE POOR MAN'S JIG BORER

The Poor Man's Jig Borer (PMJB) is a workholding device which permits the drilling of very precisely located holes with a drill press. The general idea is shown in Fig. 1 and Fig.1a (next page). Working drawings are given in Figs. 2 thru 8.

A workpiece can be placed on the Bedplate, and clamped in place with two Clamp Screws. A suitable sized Master Bushing Disk is placed on top of the workpiece, spaced away from the Fences with gage blocks if necessary, to position the Disk's center correctly for the hole location desired; the Master Bushing Disk is then clamped in place with a third clamp screw, which is somewhat longer than the other clamp screws so that it can reach across the workpiece.

A Drill Bushing the size of the hole to be drilled is put into the Master Bushing; the hole is drilled through this Drill Bushing, at exactly the desired location. Gage blocks, or slips of steel of known thickness, are used to space the Master Bushing (or the job) any desired distance from either fence. Different sizes of Master Bushing Disks and Drill Bushings can be made/bought as desired.

Obviously the accuracy of work done on the PMJB will depend upon the accuracy with which the device itself is made.

If all components are accurately made, the PMJB will enable hole location much superior to that achieved with scriber and centerpunch, certainly as good as by co-ordinate drilling on a milling machine, and possibly better. Accuracy of plus/minus half a thou should be easily within the grasp of any user, assuming the device itself is well made. And all this on a drill press!

The device I saw was made by or for a friend of mine, Mr. Kilgore Shives, when he was employed as a tool and die maker by Boeing Aircraft here in Vancouver during WWII. (It does not take a lot of imagination to recognize the practical value an item like this would have had then, with machine tools in short supply.) Mr. Shives saw the idea in the back of a booklet entitled "Metal Stamping in Small Lots", by The Dayton Rogers Manufacturing Co. of Minneapolis, MN.

Fig. 1 The General Idea

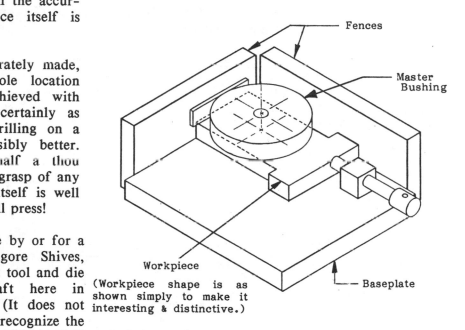

(Workpiece shape is as shown simply to make it interesting & distinctive.)

At that time, Dayton Rogers made and sold this item as a "Combination Angle Plate-Hole Locator". They made two sizes, 5-7/8 x 6-7/8" and 9-1/4 x 11-1/2"; these weighed 36 and 42 lbs. respectively. They were supplied in a wooden box with 3 Master Bushing Disks, one set of 3/8 x 1/4" unhardened parallels, 4 adjustable gaging blocks (2 for positioning the workpiece, 2 for positioning the Master Bushing Disks), 6 standard Drill Bushings in sizes 1/16, 1/8, 3/16, 1/4, 3/8 and 1/2", and a center punch.

Dayton Rogers is still in business today. I wrote, asking if they still make this item, but they do not.

The device, which I have dubbed The Poor Man's Jig Borer, can be used flat down or set up on edge on either Fence. It can be used on the surface plate, or on the table of either a drill press or milling machine.

A variation on this item might be useful to anyone involved in clock making. For example, exactly matching holes in the front and back plates of a skeleton clock frame could be located and drilled off with unerring accuracy, provided the two frame plates have been worked over to have two adjacent "reference edges" made square and true. The Fences could be shallower, while the Baseplate, which would probably want to be larger in at least one direction, could be made of aluminum to cut down on weight.

I showed Kilgore Shives' version to my friend Bill Fenton. He looked it over, and pronounced it good. He suggested an improvement, to wit, that the tapped holes for the Clamping Screws be angled down just a couple of degrees or so, so that they would tend to force the workpiece down onto the bedplate - workpiece rise is the ubiquitous bugbear of sideways clamping operations, and is commonly attacked with a lead hammer while tightening the vise or whatever.

Bill also decided he would make a rougher version for himself, for dealing with castings and odd shaped jobs. If Bill liked it, well, "What need of more words?"

NOTE: Concerning the grid of holes in the Baseplate, the several holes in the corner where the two fences meet are not needed - the one in the very corner could not be used, and at least the next nearest 3 are never likely to be used.

Note also that the fences do not meet at their common corner: this provides a relief groove there, for the same reasons such a groove is wanted in the bottom of a V-block.

One final comment: The drawings give little detail of the Shoe. The view of this part in Fig. 6 is a plan view. The Shoe's purpose is simply to provide a 2-point contact element between the end of the Bushing Clamp Screw and the Master Bushing. Make the Shoe about 1 x 7/16 x 3/8", and relieve the middle of the 1 x 7/16 face so that it will fit against the Master Bushing Disks as shown in the upper drawing of Fig. 1a.

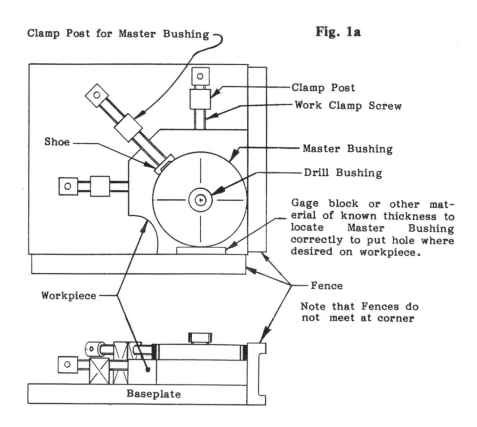

Clamp Post for Master Bushing

Fig. 1a

Clamp Post
Work Clamp Screw
Shoe
Master Bushing
Drill Bushing
Gage block or other material of known thickness to locate Master Bushing correctly to put hole where desired on workpiece.
Fence
Note that Fences do not meet at corner
Workpiece
Baseplate

Fig. 2 Baseplate

5/8" thick, surface grind
flat and square all over

Section through Fence

Fig. 3 Fences

Surface grind flat
and square all over

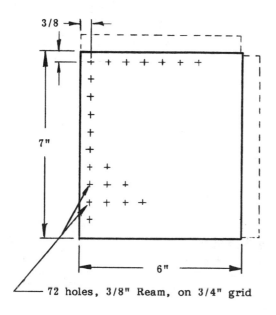

— 72 holes, 3/8" Ream, on 3/4" grid

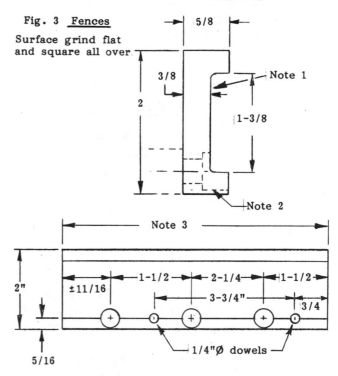

Note 1: Center portion of fences are relieved, for lightness and other reasons which will be obvious to anyone who decides to make one of these devices.

Note 2: Counterbore for 5/16" NC socket head cap screw.

Note 3: Make 2 fences, one 5.95" OAL, one 6.87" OAL

Note 4: Dimensions shown are for shorter Fence: longer Fence has 4 screws at about 1-1/2" centers, plus 2 dowels. (Note: make sure screws/dowels do not foul grid holes.)

Fig. 4 Clamp Post for Workpiece (2 off)

make from 5/8" sq. CRS

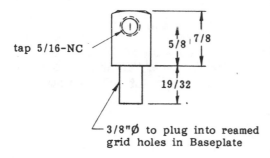

tap 5/16-NC

— 3/8"∅ to plug into reamed grid holes in Baseplate

Fig. 5 Work Clamp Screw (2 off)

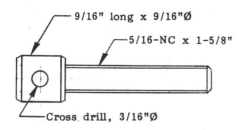

— 9/16" long x 9/16"∅

— 5/16-NC x 1-5/8"

— Cross drill, 3/16"∅

Fig. 6 Clamp Post for Master Bushings, Screw, and Shoe

make from 5/8" sq. CRS

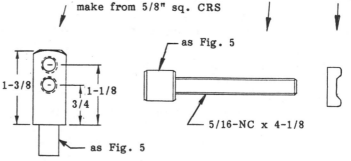

as Fig. 5

5/16-NC x 4-1/8

as Fig. 5

Fig. 7 Master Bushing (Not drawn)

Make 3, one each @ 1", 2", & 3"∅; harden and grind to exact nominal diameter. Grind/lap 5/8" hole at center. Thickness: 1/2".

Fig. 8 <u>Drill Bushings</u>

(Make one for each hole size required)

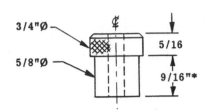

<u>Suggestion</u>: Consider using commercially made drill bushings: these are readily available from tooling component and industrial supply houses, and can be had to suit any size drill you might ever want to use.

*I'm not sure why this portion of the Drill Bushings were made longer than the thickness of the Master Bushings. It may be that what was wanted was that the Drill Bushing head would not touch the Master Bushing.

A SMALL PANTOGRAPH ENGRAVING MACHINE
(Adapted, with permission, from an article by Mac Mackintosh,
which appeared in *Model Engineer Magazine* of October 2, 1970, page 958.)

It is my experience that even the best of number and letter stamp sets intended for hand held use are unlikely to have the characters perfectly centered* on their shanks. If the stamps are then used in a fixture intended to give uniform alignment of characters in the work, they will not do so unless the shanks of the offending pieces are first corrected by surface grinding and testing of each stamp until every piece in the set "shoots into the same hole".

*(And why should they? - they are intended for hand-held use, not perfection.)

However, even this does not solve all the problems, because spacing and depth of impressions also must be correct and uniform for finished work of the best appearance. Spacing aids also can be built into the fixture, but depth is difficult to control, because blows of varying force are required to produce characters of uniform depth. A "1", for instance, requires a considerably lighter blow than a "9". All in all, it is difficult to make a first class job with stamps if more than two or three characters are involved.

If work of a better grade is wanted, the characters must be engraved. Commercial engraving machines can cost as much as a good lathe, and few hsm's could justify the expense unless planning to take in outside work, and one would need a lot of that to make it pay. I don't think that's what the average hsm wants.

Some time ago I came across a letter in a 1972 issue of M.E. The writer had built a small pantograph engraving machine to drawings published a couple of years previously in M.E. He said it was proving to be a great little rig; a photo of his machine appeared with his letter. His comments motivated me to look up the article he'd worked from - lo and behold its author was "Mac" Mackintosh, whose bandsaw blade division plate idea appears in **TMBR**. I wrote to Mac and said I'd like to put his little engraving machine in **TMBR#2**. He wrote back and said to do so, by all means, if I wanted to. So here it is - the drawings are mine, and the text here is a complete re-write based on the original. Hope you like it.

Mac got the idea for his machine from seeing a small commercial unit which produced, from master characters, 5 sizes of work from 1/2 down to 1/4 the size of the master type.

Parenthetically, while working on the re-write of Mac's article, I went to see the engraving equipment used by a stamp and sign works here. The chap who showed me around pointed out a small engraving machine they were about to ship to a customer who wanted to make his own signs. I poked my nose into the open top of the box, and there was the commercial twin to the design shown here! The machine was made by the Green Instrument Co. Inc., 295 Vassar Street, Cambridge, Mass 02139, USA.

Master type in sets, (cutters, too) are available from the Green Instrument Company, and cost on the order of $130* (for a set of 144 letters and numbers, including 3 of each numeral, plus an assortment of punctuation marks), compared to in excess of $1000 if buying the whole machine.

*May be more, or less, depending on typeface wanted, size, etc. - the price given is typical, as at mid-1988, for those typefaces you are likely to choose.

(NOTE: The operating manual for the commercial version of this machine is very comprehensive, and can be obtained separately from the Green Instrument Co. If you decide to make yourself one of these engraving machines, get a copy at an early stage in the project: when you write to them for pricing and info on master type sets, enclose US$6, and ask them to also send you a copy of "the Operating Manual for the Green Model 106 Engraving Machine.")

A good size of master type to buy for this machine would be 1/2". From this, the machine will engrave characters 1/4", 3/16", and 1/8" tall. As noted above, the commercially made machine can produce five sizes of work from one size of master type. For the sake of simplicity, Mac designed his machine to produce three sizes of work from one size of master type.

When the machine is completed, one of the first jobs for it can be to make a set of 3/16" master type, from which one can engrave 3/32", 9/128", and 3/64" work, so giving almost the same capabilities as the commercial machine for only the cost of one's time.

NOTE: part way through doing the drawings you see here, I was puzzled by a number of things in the M.E. article's drawings, so I wrote to Mac. The drawings reflect a few changes I made based on Mac's reply. Specifically, I substituted a split cotter for the split-for-clamping arrangement Mac had used in Part #4 for clamping this part to the Pillar. (I retained his split-for-clamping arrangement on the two Slide Rods which also pass through Part #4 - you may wish to re-design for cotters here also.) I changed Part #7 somewhat, and simplified the profile of Part #9 slightly. Other than that, the machine is pretty much as Mac showed it.

However, Mac also said he felt that a machine built somewhat larger all around would be an improvement. Give some thought to this before plunging into the job of making one, and *if you think the range of work you are likely to tackle might be beyond the capacity of the machine shown here, re-design it at a larger size.*

Constructional Notes

The Base (Part #1) is a piece of aluminum plate 1" thick. Lighter material would be ok, but the weight prevents the machine from moving around when it is in use.

The Pillar Bracket (Part #2) is made from a 1" x 2" block of steel or aluminum; probably you will cut this (and some other pieces - see below) from the same piece of plate used for the Baseplate. First drill and ream (or bore) the 1" hole for the Pillar, then saw roughly to shape. Machine to finished dimensions, and drill and counterbore two clearance holes for 3/8-16 socket cap screws.

The Pillar (Part #3) is not drawn, but is called as a note on the same page as the GA drawing. It can be made from 1"Ø CRS - drill rod is needlessly precise and expensive. The Pillar should not be less than 3/4"Ø, or the rigidity of the machine, and hence the quality of the engraving done, will suffer.

The Pillar is locked in the Pillar Bracket by a 3/8" setscrew. Mill a flat on the Pillar for this setscrew to bear upon. This precludes the set screw scarring the surface of the Pillar, and avoids future remorse if one desires to dismantle the machine.

> Alternative idea: Make the Baseplate a little longer from front to back. Turn a 3/4"Ø portion 0.98" long at one end of the Pillar, and omit the little flat on the side of the Pillar, but drill and tap a 3/8-NC hole about 1" deep in this end. Make a heavy machined washer about 1.75"Ø x 5/8" thick. Bore the Pillar Bracket 3/4"Ø, and suck the Pillar into place with a 3/8 x 1-1/4" soc. cap screw. This will require the Baseplate to be given some little feet, or mounted on a plywood board with a recess to take the heavy washer/screw head. This is how I think I will make mine. GBL)

Next, make the Sliding Bracket, Part #4, which carries the Slide Rods (Part #5, 2 off). This can be made from a 2" x 2" x 4" piece of aluminum. Mark out, and drill/ream a hole for the Cotter. Make up the cotter blank, and anchor it in its hole with Loctite #609. Then drill and bore the 1" hole to a close fit on the Pillar. Finish off the Cotter per drawing note.

Next, the two Slide Rod holes are drilled 7/16" - do them both at one setting in the mill to be sure they come out strictly parallel to each other and at right angles to the 1" hole. Once the holes are thus located - although not yet to their final size of 1/2" - the rest of the work on this part can be laid out, sawn approximately to shape, and finished in the mill. Don't proceed to final finishing of this part just yet.

> NOTE: As noted earlier, you might also want to consider using two more split (or solid) cotters in the Sliding Bracket to clamp up on the Slide Rods, instead of the "split for clamping" design shown. If so, see pages 92-96 of **TMBR** for info on designing split cotters, and work up the necessary details via a large scale drawing.

Make the Slide Rods (Part #5) next. (Like the Pillar, these are not drawn, but are fully spec'd in a note on the same page as the GA drawing.) Make from 1/2" drill rod - and here, drill rod is to be preferred over CRS - and on each, mill a small flat at the end, to take the setscrews in Part 6, for the same reason as on the Pillar. (Or, make similarly to the alternate I mentioned above for the bottom end of the Pillar. GBL)

The Swivel Bracket, Part #6, is made from a 1" x 1-1/2" block of aluminum. Machine to dimension, then clamp to the Slide Rod Bracket for match drilling of the two 7/16" holes, which will be opened up to their finished size of 1/2" in the next sentence. Leave the two parts clamped together and enlarge the holes to 1/2" by means of a boring head or machine reamer, to a good fit on the Slide Rods.

You can then finish the Slide Rod Bracket (Part #4). Assuming you did not re-design for cotters on the Slide Rod holes, split as drawing with a slitting saw, and drill and tap for two 3/8-16 socket cap screws to clamp the Slide Rods in place.

Back to Part #6, the Swivel Bracket: First, drill and ream a 3/16" hole through both sides, and at the same setup, enlarge the outer portion of the 3/16 hole with a #7 drill, and tap 1/4-28. Un-vise, flip, vise up again, pick up the other end of the 3/16 reamed hole, and repeat the 1/4-28 routine. Put a 1/4-28 set screw in each of these two holes, insert the Slide Rods, tighten the set screws onto their flats thereon, and slide Part #4 onto the Slide Rods: the Rods should work smoothly, without any tendency to bind, over the full range of possible travel, and should lock solidly in position when the cap screws in #4 are tightened.

Make the Swivel, Part #7, next. Machine to dimensions shown, so that it fits the notch in the Swivel Bracket (Part #6) without side play. Drill and ream the 3/16 hole, and cut a piece of 3/16" drill rod to a length of 1.48" long, to fit into the Swivel Bracket when the Slide Rods are in place. Bore a 0.500"Ø flat bottomed hole, 5/8" deep, in the top of the Swivel, and into this Loctite a 1/2"Ø steel insert tapped 1/4-28. (This is the "18th bushing", per note re Bushings on the drawings. GBL)

General Arrangement Drawing
for a
Small Pantograph Engraving Machine

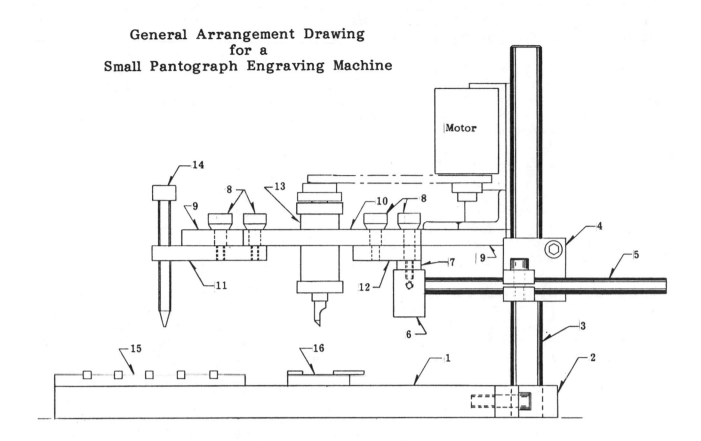

Part #1 Base

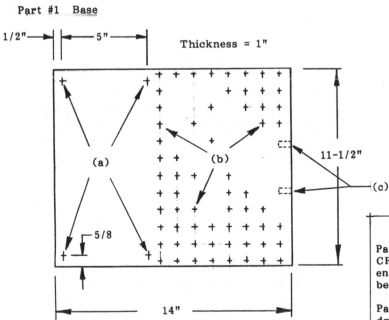

1/2" ◄► │◄— 5" —►│   Thickness = 1"

11-1/2"

14"

(a) 4 holes, tapped 1/4-28, to hold Part #15

(b) 80 holes, on 1" x 1" grid, tapped
    1/4-28 to take work clamps, etc.

   NOTE: Center this pattern of holes on
   the 8" x 11-1/2" area not occupied by
   Part #15. In this drwg, these holes are
   biased a little too far towards the area
   occupied by Part #15.

(c) 2 holes, tap 3/8-24 x 3/4" deep,
    to take Part #2

NOT DRAWN:

Part #3, Pillar - Make one, from 1"Ø
CRS, 11" long, with a flat milled at one
end for 3/8-24 setscrew in Part #2 to
bear on.

Part #5, Slide Rods - Make 2, from 1/2"Ø
drill rod, 9" long, with a flat at one end
for 1/4-28 setscrew in Part #6 to bear on.

Bushings - Make 18, 17 for Arms, and
one for Part #7. See text.

We now come to a chicken/egg situation - which to describe first, the Arms, or the Bushings that are put into the several 1/2" reamed holes in the Arms? Let's look at the Bushings first, and then deal with the Arms they go into.

The Bushings are not drawn, but are called out as a note on the drawing. First, a comment: The bearings in the arm joints of a pantograph engraving machine are critical to its performance, and must be very nicely fitted. On larger commercial machines, high class ball bearings are used. On our little machine, bush type bearings, press fitted into the aluminum arms, are used. This form of construction is entirely appropriate for a small machine such as this.

From 1/2" brass rod, part off 17 pieces, each 17/32" long, and then face each one to .500" long, ± 0.005".

Eight of these are then drilled and reamed 3/8", eight are drilled and tapped 1/4-28, while the last is drilled and tapped 3/8-40. (Lacking a tap of this size, you could make one, or screw-cut both the internal and external threads, or break down and buy a tap. You could also use a 3/8-24 thread, but the finer pitch is to be preferred, for reasons given later.)

Next, from 3/8"Ø drill rod, turn up four Pivot Bolts, Part #8. The knurled heads - brass, mild steel, or aluminum - are pressed or Loctited on, after the business end of each is threaded 1/4-28 as drawing. The 3/8"Ø bearing portion must be carefully fitted to the reamed bushings so that there is approximately 0.0003" radial clearance* and 0.001" axial clearance. Thus fitted, the joints will be free to move, but will be without appreciable shake. Three of these Pivot Bolts are made long enough to accommodate one 3/8"Ø reamed bushing (1/2" long), while the fourth is longer, to take two of the bushings, one on top of the other, at the Swivel, Part #7.

> *NOTE: If the bushings are reamed and then pressed into an undersized hole, the I.D. of the reamed bush will close down slightly. Therefore, you will probably need to ream the bushes, press them into place, and then put the reamer through again to take care of any "squashing" that has occurred.
>
> For myself, I think I will do things a little differently. I would make the pins from 3/8 drill rod, after which I would rough bore the bush undersize, press the bush into the arm, and then, using an expansion reamer, ream the bushes to a nice fit on the pins, or bore to finished size with my boring head. If I had to finish the bushes with a regular (non-adjustable) reamer, I would first make up a test piece, and see what sort of fit I got before I started on my pantograph parts. Alternatively, I would lap the bushes, either by hand or on a pin hone machine - I have a friend who owns one. To find someone in your area who has one of these wonderful little machines, look in the Yellow Pages under "automotive machine shops".
>
> Or - why not make life simple? - bore the bushing holes 0.501"Ø, and Loctite the bushes into same - they will not come out, and if the as-reamed hole is a nice fit on the 3/8" drill rod Pivot Bolt, it will be the same size when installed as before the bush was put into its hole in the Arm. GBL

(Another point to consider: One must keep in mind that the engraving machine is really a small milling machine. As such, it wants to be quite rigid. I think it might be a good idea to lengthen the Pivot Bolts slightly, and put Teflon washers in the various elbow joints of the Arms. Otherwise you will have aluminum rubbing on aluminum, and this cannot be ideal. GBL)

The Pantograph Arms, Parts 9, 10, 11 and 12, are made out of 1/2" aluminum plate. All pivot holes are drilled and reamed .4985", so that the bushings will be a press fit in them. (Or, see comments above.)

In Part #9, the big hole in the middle of the Arm has a 1.25" x 40tpi internal thread to take the Cutter Spindle Casing (Part #13), which has a similar thread cut on its O.D. If your lathe is big enough to swing the Arm, you can screwcut this thread directly in the Arm itself. If you can't

Part #2 Pillar Bracket

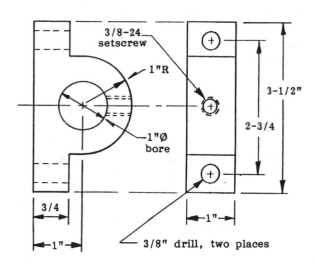

3/8-24 setscrew

1"R

1"Ø bore

3-1/2"

2-3/4

3/4

1"

1"

3/8" drill, two places

Part #7 Swivel

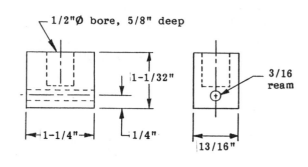

1/2"Ø bore, 5/8" deep

1-1/32"

1-1/4"

1/4"

3/16 ream

13/16"

Part #6 Swivel Bracket

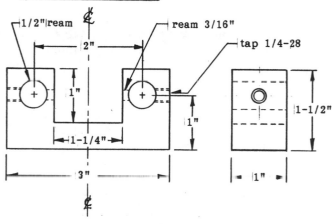

1/2"ream

ream 3/16"

tap 1/4-28

2"

1"

1"

1-1/4"

3"

1-1/2"

1"

Part #4 Slide Rod Bracket

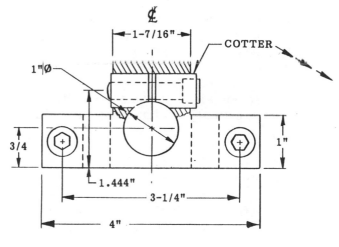

1-7/16"

COTTER

1"Ø

3/4

1"

1.444"

3-1/4"

4"

COTTER: 5/8"Ø x 1.575" long; counterbore end of cotter 0.515"Ø x 0.260" deep, drill & tap for 5/16" x 1-3/8" NF soc. cap screw. Loctite cotter blank in place in a 5/8" reamed hole in the Slide Rod Bracket. Bore the 1"Ø hole for the Pillar. Then, heat parts to break Loctite bond, remove cotter, saw in half to activate clamping action, face sawn ends, and bevel outer ends.

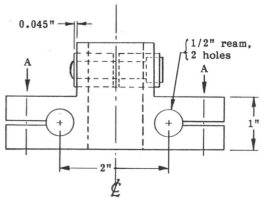

0.045"

A

1/2" ream, 2 holes

A

2"

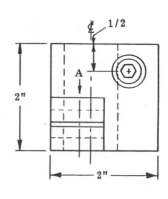

1/2

A

2"

1"

2"

Re arrows labeled "A" in the drwg at left: The Slide Rod Bracket is split for clamping of the Slide Rods, which go in the 1/2" reamed holes in the Slide Rod Bracket. Fit 3/8 x 1" NF soc. cap screws at the points indicated by arrows A.

swing the Arm in your lathe, do as drawing: use your vertical mill to bore a 1-3/8"Ø hole, and into it Loctite a flanged steel bushing (Part #9a), which latter item is easily chucked and internally screwcut 1.25" x 40tpi.

NOTE - Something else not shown on the drawings: Mac's machine had a split about 1-1/4" long centered in Part #9, just where I have put the dimension "1-5/8" on my drawing. He then fitted a 10-32 screw across the Arm, close to the big threaded hole. This permitted him to screw the Cutter Spindle Casing up or down as required, and then lock it at that position with the cross screw. This feature is not shown on my drawing of Part #9. There are at least three ways to achieve this lockup capability:

    1. Loctite the bush in place, and then saw a slot just as did Mac, just as if the bush was part of the Arm.

    2. Make a 1-1/4-40 ring nut, split it, fit a tangential adjusting screw, and install this ring nut on the threaded body of the Spindle Housing, above or below the Arm.

    3. Leave more metal in one of the 45° flats around the big threaded hole, and fit a padded set screw.

Some notes on making Part #10, the Offset Arm: Most of this Part is "in the wind" - i.e. not critical for size, etc. How you go about making it will depend on various factors, but here's the procedure I will likely use, and I have put the necessary extra* info on the drawing so you can do likewise.

From 1/2" aluminum plate, saw out a piece of material about 1-5/8" x 7". Mill two adjacent edges square to each other - the dotted lines at the right hand side of the drawing indicate these two good edges; the other two sides can be left as they come from the hacksaw, except for burrs.

Blue the surface of the material, set it on the surface plate on its good long edge (= "e"), and scribe lines full length at heights of 3/4", 15/16", and 1-1/8" above the surface plate. Stand the material on its good end, and scribe 4 more lines at heights of 3/8", 1.358" (= 3/8 + 0.983"), 5.392" (= 1.358 + 4.034") and 6-3/8" above the surface plate.

Clamp down on mill table, with the long machined edge "e" parallel with mill table axis. Pick up the first bushing hole center "a", and proceed to drill/ream the two 1/2" dia. holes "a" & "b" 6.000" apart.

Leaving the job undisturbed, drill the two 3/8" dia. holes "c" & "d" next. (Note: I'd be strongly inclined to drill these two holes 1/64" oversize (i.e. 25/64") so my fly cutter could later run out into thin air in the holes rather than having to come out exactly tangent to the wall of the hole - why make a job like this any fussier than it needs to be?) After drilling these holes, hacksaw out most of the excess material along sides "j", "k", & "f" before milling to finished shape as in next paragraph.

Bolt an angle plate to the mill table and clamp the job to same with edge "f" in the clear - i.e. looking up at the mill spindle nose - and mill this edge to dimension. Reset job at 45° and mill edges "g", "h", "j", and "k", (4 set-ups; note letter "i" is not used in the sequence)

I'll probably file the two outer corners to profile by eye, and file the end profiles with the aid of a pair of filing buttons.

    * If you want to use the approach outlined above, notice that your layout work is reduced to the scribing of 8 lines. My drawing provides all the info needed; if the 0.983" and 4.034" dimensions were not given, you'd have been scratching your mathematical navel for a while, working them out for yourself.

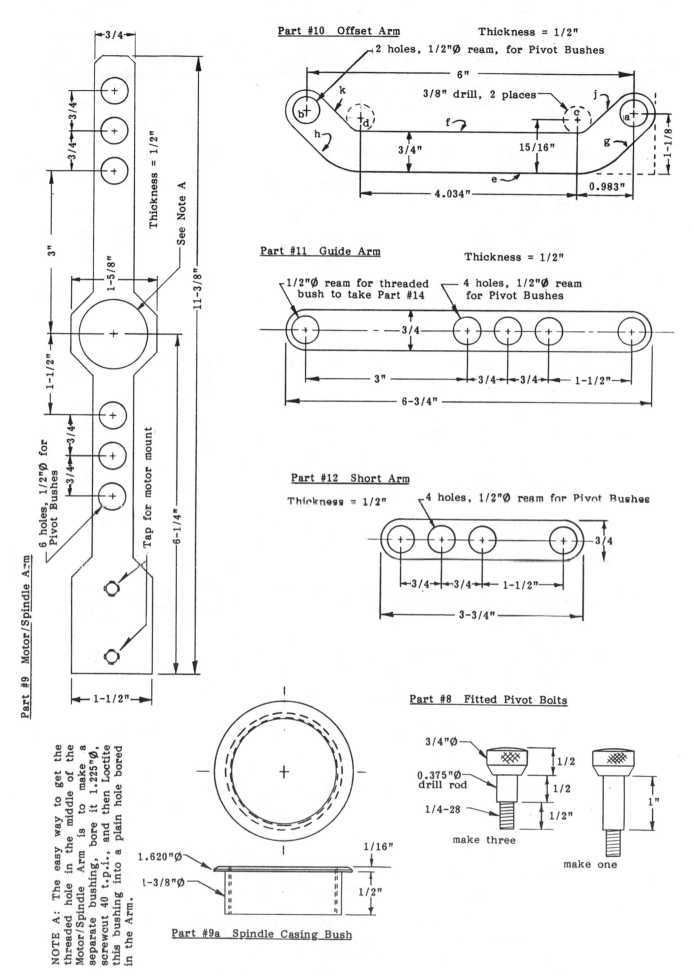

Part #10 Offset Arm Thickness = 1/2"

2 holes, 1/2"Ø ream, for Pivot Bushes

6"

3/8" drill, 2 places

k j

b d f c a

h 3/4" 15/16" g

e 1-1/8"

4.034" 0.983"

Part #11 Guide Arm Thickness = 1/2"

1/2"Ø ream for threaded bush to take Part #14

4 holes, 1/2"Ø ream for Pivot Bushes

3/4

3" 3/4 3/4 1-1/2"

6-3/4"

Part #12 Short Arm

Thickness = 1/2"

4 holes, 1/2"Ø ream for Pivot Bushes

3/4

3/4 3/4 1-1/2"

3-3/4"

Part #8 Fitted Pivot Bolts

3/4"Ø 1/2

0.375"Ø drill rod 1/2

1/4-28 1/2

make three

1"

make one

Part #9 Motor/Spindle Arm

3/4

3/4 3/4

3"

Thickness = 1/2"

1-5/8"

See Note A

11-3/8"

1-1/2"

6 holes, 1/2"Ø for Pivot Bushes

3/4 3/4

Tap for motor mount

6-1/4"

1-1/2"

NOTE A: The easy way to get the threaded hole in the middle of the Motor/Spindle Arm is to make a separate bushing, bore it 1.225"Ø, screwcut 40 t.p.i., and then Loctite this bushing into a plain hole bored in the Arm.

1.620"Ø

1-3/8"Ø

1/16"

1/2"

Part #9a Spindle Casing Bush

31

The Cutter Spindle Casing, Part #13, is turned from 1-1/4"Ø mild steel rod, and the O.D. screwcut 1.25" x 40tpi full length. It is bored 3/4"Ø right through and recessed at each end to take 1/2" I.D. x 1-1/8" O.D. ball bearings - I would suggest a pair of New Departure R8 single row deep groove Conrad ball bearings or equal; the equivalent SKF part number is EE4. These are rated to 25,000 rpm and can take radial loads on the order of 750 lbs. A good figure to use for the thrust load capacity of this type of bearing is 50% of the radial load rating.

Trivia: the single row deep groove Conrad ball bearing was designed about 1888, and so far as its geometry is concerned, it has never been improved upon. Materials, speed and load capacities, etc., are higher now than originally, yes, but they got the basic design right the first time.

The spindle that carries the actual engraving cutters takes some care in making, and some thought beforehand as to design.

Make the Cutter Spindle from 5/8"Ø drill rod. In machining this part, get the axial location of the shoulders for the ball bearings quite accurate, so that in the assembled Spindle, longitudinal play can be reduced to zero.

This is accomplished by making the lower end of the Cutter Spindle a light press fit in the inner race of the lower bearing, while the upper end is made a tight slide fit in its bearing race. With this arrangement, any end play that may be present can be removed by tightening the Bearing Cover Caps.

NOTE TO READER: When Mac made his machine, he had on hand some cutters as used in the commercial machine, and he made his spindle to accommodate them. The next 3 paragraphs are the text of his own article, virtually verbatim. Although few if any who make this machine from this presentation will have reason or need to do likewise, the procedure he describes does point up the need for accuracy in this part of the project:

"Preliminary setting of the top-slide for the taper was done by cutting two dummy tapers in some scrap; the spindle was then carefully centered, bored for the drawbar and a satisfactory taper (so far as fit was concerned) was obtained.

Unfortunately, when mounted in the bearings, it was found that there was a runout of 0.007" at the tool point. This would have made any engraving rather sloppy, so this spindle was discarded.

A second attempt showed a runout of less than 0.001". I had anticipated that a second attempt might be necessary and had not upset the setting of the top-slide before testing the spindle, so the machining of the second (& successful) spindle took much less time than for the first (unsuccessful) one."

A drawbar is made from 3/16 drill rod, and fitted with a knurled brass knob at one end, the other being screw-cut 10-32 to match the engraving cutters.

NOTE: If you decide to make your own cutters, use a self releasing taper for the cutter shank. The drawbar will hold it in place in use, and it will drop out easily when you want it out.

A sewing machine motor will give plenty of power for driving the cutter spindle. Two small three-step pulleys are required, one bored to fit the motor spindle, and the other for the end of the Cutter Spindle. Appropriate cutter speeds for the small cutters used are on the order of 5000 to 9000 rpm. Size your pulleys accordingly, based on the speed of the motor you use.

The Motor Bracket is shown only schematically, in the GA drawing - make to suit the motor used. Slots should be provided to permit taking up slack in the belt, and also for vertical positioning of the motor with respect to the main spindle pulley.

Plan view of Arm Assembly

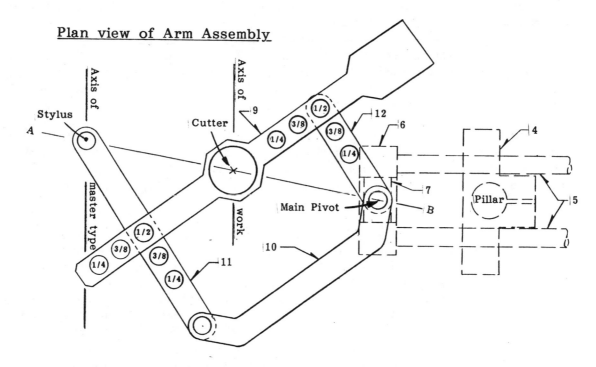

The line A-B is interesting: The distance along it from main pivot to cutter divided by the distance from stylus to main pivot governs the size of engraving that will be produced from a given size of master type. As Part 9 is moved closer to Part 10, and at the same time rearward (toward the upper right corner of this page) along its own axis, smaller work will be produced from the same master type.

Part #13 Spindle Assembly

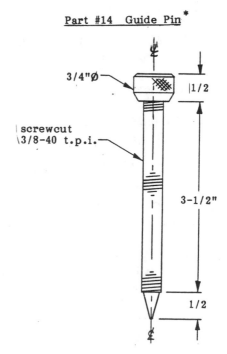

Cutter Spindle, 1/2"Ø x 1-1/4" this portion; drill #7 (= 0.201") full length.

End caps, 1-1/2" O.D. x 0.4" thick, bored 1.225"Ø x 1/4" deep, and screwcut 40 tpi.

Spindle Casing, 1-1/4"Ø, 2-5/8" long, drilled 3/4"Ø full length, and screwcut O.D. 40 tpi full length.

Cutter Spindle, 5/8"Ø x 2-1/8 this portion.

Bearings, 1/2" I.D., 1-1/8" O.D., 1/4" thick; see text.

End cap, as above

Taper socket, to suit cutters

Part #14 Guide Pin*

3/4"Ø

screwcut 3/8-40 t.p.i.

1/2

3-1/2"

1/2

*This part is termed the "Stylus" in the drwg at the top of this page.

1/4" round plastic belting, obtainable from "Small Parts, Inc." will be found satisfactory for applications such as this. (O-ring material is also used for such jobs, although whether with equal or lesser success I do not know. GBL)

The last item required for the engraving machine itself is the Guide Pin, or Stylus, Part #14. Make this from 3/8" CRS threaded 3/8-40 (or 3/8-24) for its full length. Turn a nicely tapered and polished round ended point at one end to fit the master type. Loctite a brass knurled knob at the top end, and make a knurled Locknut (not drawn).

The use of 40 tpi threads on both the Guide Pin and the Spindle Casing make setting the pantograph for engraving depth an easy matter. The two threads do not have to be of like pitch, but when I queried Mac on this point in a letter, he replied thus: "Not necessary, but convenient. You can give a half-turn to each of them and get the depth to 12-1/2 thou if you stick to 40 t.p.i."

The Rack for the master type, Part #15, is made as drawing, to take 12" lengths of 1/4" sq. CRS, the latter anchored in place with three 4-40 socket cap screws per piece. Space the slots at intervals to suit the (purchased) master type. Typically, these will have been made from 3/4" x 3/32" brass strip. The slots should be a close fit on the master type, with not more than 0.005" clearance. I asked Mac if somewhat less clearance would be preferable. He replied: "....remember that error is reduced (in the work done); 0.005" is quite ok, and lets the master type slide easily in the rack."

The rows of 10-32 holes on 1" centers are for clamps and stops for locating the master type.

Make a couple of slotted slides to fit the built-up slots in the Master Type Rack. These slides are held by knurled screws, and hold the master type against lateral movement in the Rack.

You will soon accumulate an assortment of shop-made fittings for holding workpiece blanks of various shape. One example of this sort of fitting may be worth describing. See Part #16. This item is similar to the Master Type Rack, but has slotted mounting holes and an adjustable slide to accommodate material of variable width for engraving.

If engraving larger than 1/4" is wanted, it would be necessary to purchase a larger size of master type. In this case one would do well to first draw up a machine larger in all particulars, because considerably more material is taken out in engraving a character 1/2" tall as opposed to one 1/4" tall, and the machine should be beefed up all around for work of this size.

The machine shown should suit the average hsm very nicely - it is basic, small, and simple, and should satisfy most needs. It "has everything" without the need for the builder to design additional elements to be able to do "builder's plates", club member name badges, etc.

If you want to engrave numbers on dials, you will need to raise the Master Type Rack by means of a riser block. Then, mount the dial on an angleplate, and clamp the latter down between the Pillar and the masters.

IDEA: One could give some thought to making the engraving machine as a detachable accessory for the vertical milling machine or (perhaps better) the lathe. It could be set on the lathe bed and a pulley on a stub axle could be stuffed into the 3-jaw chuck. With a spring-loaded jockey pulley arm, the lathe could drive it. Mac says he does not favor this approach, as it is more convenient if the engraver is a separate self-contained machine, usable without disrupting the other work of the shop.

Accessories for accommodating other types of work can be devised by the builder, either along lines shown in the literature available from the makers of commercial engraving machines, or from one's own thinking.

Part #15 Master Type Rack

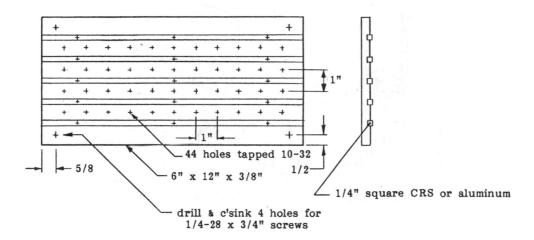

1"

44 holes tapped 10-32

6" x 12" x 3/8"

5/8

1/2

1/4" square CRS or aluminum

drill & c'sink 4 holes for
1/4-28 x 3/4" screws

Part #16 Rack for Work of Variable Widths

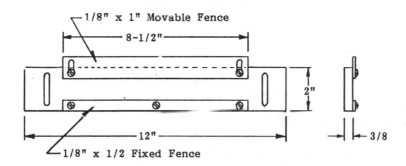

1/8" x 1" Movable Fence

8-1/2"

2"

12"

3/8

1/8" x 1/2 Fixed Fence

NOTE: Mac's article shows almost nothing about the nature of cutters to use with this machine. On the following page I show, in general terms, how the business end of such cutters would want to be made. Home made engraving cutters of this type can be sharpened on a bench grinder with the TINKER Tool & Cutter Grinding Jig, or with some other grinding device especially made for the job, typically by means of a swash-plate collar fitted to the cutter grinder's toolholder, said swash-plate being arranged to give axial backing off during a 90° rotation of the cutter. There is an article in M.E. for March 3, 1955, page 232, wherein just such a device is shown; a photocopy of the article can be obtained from the publishers of *Model Engineer* by writing to them at the address given in the Appendix. There will be a charge for the photocopy and the mailing costs - I'm not sure how much. I'd send $3 or 4; I'm sure they'd inform you if this is insufficient.

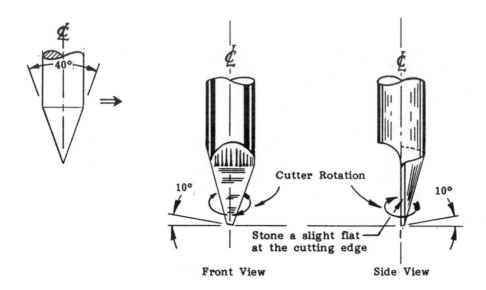

Cutter Rotation

10° Stone a slight flat — 10°
      at the cutting edge

Front View   Side View

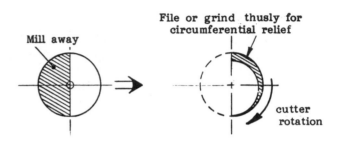

Mill away   File or grind thusly for
                   circumferential relief

                                    cutter
                                    rotation

View from above cutter

Added to the 7ᵗʰ Printing, January 1999:

The twists and turns that occur along one's path are sometimes strange indeed. One day some years before I wrote this book, I chanced to pass a shop on a quiet side street off one of the main drags in Vancouver. In the window of the shop, I saw a strange machine about the size of a bench drill press. I went in and asked the purpose of the machine. It turned out the shop was a printing shop, and the machine was used to engrave printing dies. I contacted the manufacturer, learned a few things about these machines, and after a while the matter sank back into the morass. Over the years, other ideas came to roost, and got sorted out. Then one day in 1995, a number of these old chickens came marching out in an entirely new grouping, which caused me to recall the printing shop... I phoned them: could I come down and see their engraving machine in action? Yes, but come soon, because we're about to close for good. I went down the next day, and while there, learned the reason the shop was going to close: the owner wanted to retire. To make a long story short, I bought the engraving machine, restored it to good-as-new condition, and bolted it to a choice spot on my workbench. Its capabilities would amaze you. Reduction rates of 100:1 and higher are routine business, with flawless duplication of simple hand cut masters, which I have learned to make. I am currently in the process of boiling down all this and much more info, for reincarnation in a new book and video, to be titled **The Secrets of Lettering on Metal.** Please call me at (604) 922-4909 if you would like a copy when available. **Guy Lautard**

MILLING SPINDLES AND OVERHEAD GEAR

A useful adjunct to an hsm's lathe is a "milling spindle" - sometimes referred to as a milling and drilling spindle, because usually both milling and drilling can be done with such a unit. The shorter term is more convenient.

A milling spindle increases the variety of work that can be done on the lathe. Typically it is mounted on the vertical slide, or to the cross slide or toolpost, and driven via a belt running over stepped pulleys, which give various spindle speeds. Operating speeds vary with the job, but it should be able to run at speeds up to 8,000 rpm or more.

Drilling and light milling operations can be carried out; one field where such devices are commonly used is in clockmaking, where the spindle is used to run cutters used to cut clock gears. Such equipment is obviously not for heavy milling - it simply does not have the mass and rigidity required for such work - but it does open up whole new fields. Shop-made cutters can be run at speeds suitable to their intended use, while the lathe and chuck serves as work holder and dividing head.

I have not made myself a milling spindle, so I am not able to give as much advice on the matter as I would like. I will tell you of a source of a commercially made milling spindle, and will describe in general terms the features one might incorporate in a shop-made unit. If you make one, design it to suit yourself and the materials available to you.

A milling spindle can be mounted on the toolpost, or the vertical slide, or otherwise, as suits the lathe and its user. If the milling spindle is to be toolpost mounted, it is desirable to be able to set the spindle at lathe center height automatically. This can be arranged, when making the unit, by boring the spindle housing while it is mounted on the lathe toolpost as it will be when in use.

> TIP: If you prefer to mount your milling spindle on the lathe's vertical slide, make a "setting slug" to aid in setting the milling spindle at lathe center height. The slug would be placed on the lathe bed. The vertical slide, with the milling spindle mounted thereon, would then be lowered until some datum surface contacted the top of the setting slug, at which point the milling spindle would be dead on center height.

A milling spindle should be solidly and well made, with minimum "overhang" at the business end, if good service and performance is expected. This means good bearings, free of play, and capable of sustained hi-speed operation when appropriate.

The bearings can be bush type - bought or home made - or they can be ball bearings. Bushings can be cast iron, bronze, oilite bronze, or even some of the plastic materials such as Teflon, PTFE, etc. The bearings must be able to handle axial thrust as well as radial loads. Conventional deep groove ball bearings will handle thrust loads up to about 50% of their radial load rating, hence will do fine so long as they are properly mounted. Bush type bearings can simply include an integral thrust face on the bushes at both ends of the spindle - see Fig. 1.

Most of the jobs one would use such a spindle for is relatively light duty work, hence the spindle need not be of large diameter, although it does need to be well made. It can be solid or hollow full length, with a taper socket, or other means of adapting cutters etc. to the business end.

A milling spindle in about its simplest form, is shown at Fig. 1. This can be elaborated upon in several directions:

> The spindle housing can be given refinements to enable it to be used as a dividing attachment,

> and/or it can incorporate a swivel base, with or without a 0-360° protractor scale engraved thereon.

The spindle nose can be made to duplicate the lathe spindle nose, and/or to take factory or homemade collets.

If very slow speed running is desired, the spindle can be arranged to be driven through reduction gearing, employing the lathe's change wheels, or other factory- or shop-made gears.

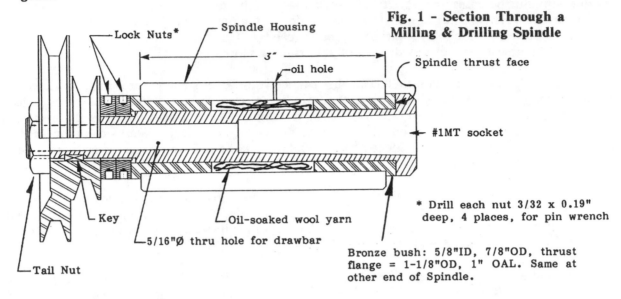

Fig. 1 - Section Through a Milling & Drilling Spindle

Lock Nuts*
Spindle Housing
3"
oil hole
Spindle thrust face
#1MT socket
Key
Oil-soaked wool yarn
5/16"Ø thru hole for drawbar
Tail Nut

* Drill each nut 3/32 x 0.19" deep, 4 places, for pin wrench

Bronze bush: 5/8"ID, 7/8"OD, thrust flange = 1-1/8"OD, 1" OAL. Same at other end of Spindle.

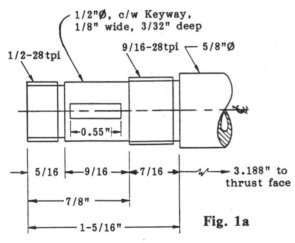

1/2"Ø, c/w Keyway, 1/8" wide, 3/32" deep
9/16-28tpi
5/8"Ø
1/2-28tpi
0.55"
5/16
9/16
7/16
3.188" to thrust face
7/8"
1-5/16"

Fig. 1a

Notes Re Milling Spindle Drawing

1. This drawing is an accurate scale drawing. Dimensions not noted can be measured and scaled.

2. The Spindle Housing as drawn is a 1-5/16" diameter piece of round material. You might choose to make the Spindle Housing from 1-3/8" square CRS. If so, and if you like the Overhead drive arrangement detailed in the section following, buy a slightly longer piece of 1-3/8" square CRS than would be required for that project alone.

3. Fig. 1a is a scrap detail of the tail end of the spindle as drawn in Fig. 1 above. All threads, and the nuts for them, should be screwcut, hence oddball specs for same.

If using bush bearings, reduce the diameter of the spindle between the bushes, and wrap some oil-soaked wool yarn around this area of the spindle before assembly. The yarn serves as an oil reservoir, providing a reserve supply of oil for the bearings, in the event thee fail to add oil when thee should.

The business end can take various forms depending on what you plan to do with the finished unit. A #1 MT socket is not a bad choice. Alternatively, a straight bored hole may suit your purposes, with or without a closing taper at the outboard end. You can make a collet type chuck as at Fig. 2, or you can put a thread or a male taper on the spindle nose to take a commercial drill chuck - if you go this route, buy the best chuck you can afford; if for drilling only, a keyless type would be best, although the best of the Jacobs key-type chucks are also good, and being shorter (= less overhang) would be preferable if light milling operations are also contemplated.

NOTE: When tightening any key operated drill chuck, use all three holes! Snug up by hand, put the key in the first hole, and tighten up somewhat, then some more via the second hole, and finally, cinch up tight on the third hole. The chuck maker didn't put three key sockets in there for your convenience - they're there so you can tighten the chuck up properly. Use 'em all.

To get the I.D. and O.D. of a workpiece (the spindle) concentric, the best procedure is normally to rough machine the O.D., leaving some material for finish machining, then drill the draw bar hole, if there's going to be one, ream or otherwise cut the female taper, plus a concentric conical register at the other end, and mount the spindle on an accurate mandrel - specially made for the purpose if need be - to finish the O.D. perfectly concentric with the I.D. Alternatively, bore the interior with the otherwise finished spindle running in its own bearings.

When completed and assembled, the Spindle should run freely, but without endplay perceptible to an indicator.

For those who want a collet system, I have drawn a collet/chuck arrangement at Fig. 2, and if you like this idea, you could scale this drawing for dimensions not given, or draw a similar collet/chuck of your own design. 6X full size is a good scale to use. You may think to yourself: "My chuck will be 1-1/4"Ø, and 1-3/8" long, because I want to be able to hold bigger shanks than 1/4"Ø max." Draw in a centerline, draw a 1-1/4" x 1-3/8" rectangle, and proceed to work up a chuck to suit yourself, all of it within that rectangle.

The collets per se should be made up in some sort of permanent tooling preserved thereafter for the production of additional collets as needed.

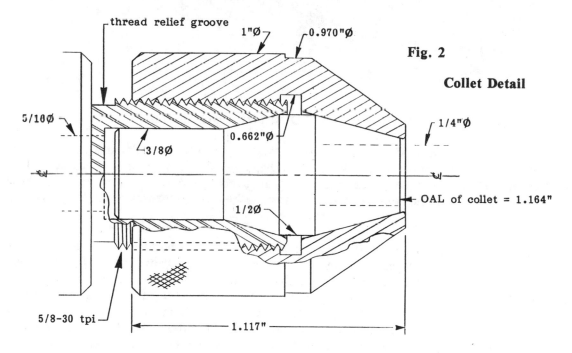

That about covers most of what I have to say about shop-made milling and drilling spindles, except for a couple of points:

There is a kit of castings available for an item of excellent repute known as the Potts Milling and Drilling Spindle. Potts was a Welshman who developed a number of products for model engineers. He was active in the 1930's and for a long while later. Castings for tools to his designs are still available. Power Model Supply shows the Potts M&DS at about $35 in the latest catalog I have; it's probably a little more by now. I have one of these kits of castings and have been told, by a couple of chaps who should know, to get busy and make it up, for it is, they say, excellent.

39

If you'd prefer to buy a ready-made ball bearing equipped spindle (as photo below), write to Arrand Engineering (see Appendix). I have one of these units, c/w flywheel type drive pulley, belting, etc., and it certainly seems to be well and solidly made. The spindle socket is #1 MT. Arrand offers a variety of tooling on #1 MT tooling shanks, as well as blank ended #1 MT arbors which the buyer can finish off as required. The basic unit was about US$100 in bald round numbers, as at mid '87 - write for a current price list.

AN OVERHEAD DRIVE FOR YOUR MILLING SPINDLE

The next consideration is a means of driving a milling/drilling spindle unit for use on the lathe. The classic arrangement is via what is known as "overhead gear" - a sort of one-machine line shafting installation. There are several ways to set up an overhead drive system, and just how you go about it will depend on the type and size of your lathe, the nature of your shop, and so on.

One way is to rig two vertical pillars, from floor (or lathe bench, or whatever) to ceiling, with a couple of horizontal braces to keep things in place. Bearings to carry a rotating shaft are placed at some convenient height. The shaft is driven either from the lathe motor or (more likely) from a separate motor. The shaft will have a keyway cut over most of its length, to drive pulleys which can be positioned anywhere along it. Another belt is run from one of possibly several drive pulleys on the overhead shaft down to the milling spindle. As the milling spindle is moved along the lathe bed, or as different combinations of pulleys are used to get appropriate cutter speeds, tension on this drive belt is maintained by means of jockey pulleys, the latter typically being mounted on a hinged or counterweighted arm.

There are 3 or 4 alternatives that come to mind.

1) Erect a pillar behind the lathe, and on it put a boom that is adjustable for up and down, and all other movements. Put a small motor on one end of the boom, and run a belt from same to jockey pulleys at the other end and from there down to your milling spindle. A nice and compact arrangement of this type was described in a 6-part serial in *M.E.* beginning in Nov. 20, 1981, page 1382; castings and drawings are available from C.E.T. Associates here in North America. I don't know price, but it is a slick arrangement employing several aluminum castings, and will not be cheap. You can write them for details if interested; address is in Appendix.

2) Attach a motor to a board connected to part of your lathe carriage, and run your drive belt from there to the milling spindle. The motor follows the carriage wherever it goes; the only thing I'd be leery of here would be putting an undue strain on a cross slide T-slot, or throwing the carriage slightly out of line - this latter need not be a problem if the saddle gibs are correctly adjusted, and if the whole thing is made with an eye to a good sound installation, rather than cobbling together anything at all, just to see the belt going around.

3) Rig the motor more or less direct to the milling spindle (not dispensing with belt drive), pretty much like a commercial tool post grinder.

4) Rig a pivoted boom from the rafters above and just behind the lathe, as in Fig. 3. Such an arrangement was described by a J. Nixon in the May 9, 1957 issue of *Model Engineer Magazine*, on page 681. Nixon said he had tried several arrangements but had settled on this particular one some 30 years previously, and had found it entirely satisfactory, its only disadvantage - if one could consider it to be such - being that it required a motor all to itself. (So would most other arrangements, and a 1/3 or 1/4 hp motor from a washing machine or similar can typically be picked up for maybe $5.) The belt can be crossed in a moment, if a reversed drive is desired. It can equally quickly be disconnected and looped up onto the Boom when not required. (As for type of belt to use, see the section on making **A Small Pantograph Engraving Machine.**)

When I was making the drawings which follow, showing Nixon's arrangement, the more I studied the design, the more I liked it. I think it is a thoroughly practical rig, and I intend to build my own "overhead" almost exactly like Nixon's.

I say "almost..." because I made a couple of small detail changes - improvements, I think - to the design as he showed it in *M.E.* For example, he showed the Boom in two pieces screwed into the Trunnion Block. What I show here is far simpler: bore a straight hole in the Trunnion Block, slide a piece of pipe long enough for the whole Boom into said hole, cross drill the Block and the Boom at the desired point, and lock them together with the 3/8" CRS pin which serves as the pivot for the Boom assembly.

If you wish to follow this design, I do not think you will be disappointed.

5) Another arrangement I have heard of is "dwarf overhead gear", wherein a belt is brought up from a motor located under the lathe, and run around various pulleys and so on - no need to detail it further - if you like the idea in preference to an overhead arrangement, it would be easy to devise.

6) Finally, there is the McDuffie Drive. This was devised by one Ted McDuffie of Florida for driving his Myford Super 7 lathe. It consists of a motor on a sliding platform running on a horizontal rod parallel to, and behind, the lathe bed. The motor can be slid to any position desired, and used to drive the lathe from either sheave of a 2-diameter pulley on the motor shaft. When the user desires, he can flip the main drive belt off, and slide the motor to any place he pleases along the slide rod, and run a belt to a milling spindle or whatever on the lathe carriage. Bill Smith, whose Grasshopper Skeleton Clock was mentioned in **TMBR**, saw Ted McDuffie's drive arrangement on a visit in late 1986, and told me about it. I wrote to Mr. McDuffie, asking for details. He replied, sent me a photo, and said the McDuffie Drive was going to be fully described in *Model Engineer* in the next few months - it may have already been published by the time TMBR#2 hits the press.

Fig. 3 – General Arrangement Drawing for a Practical "Overhead" Drive

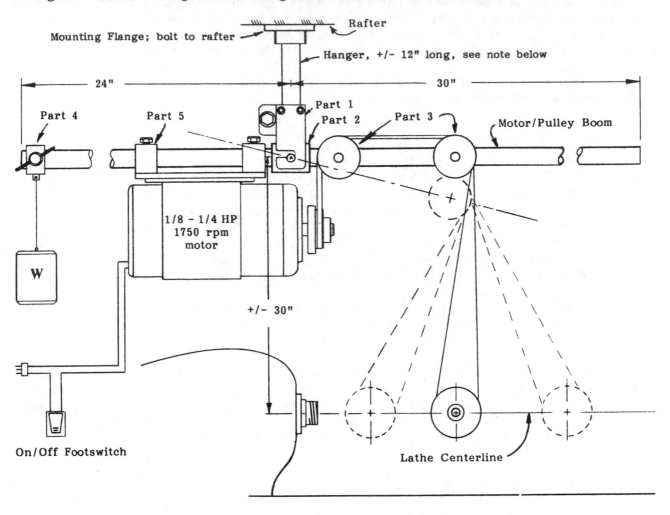

NOTES TO GENERAL ARRANGEMENT DRAWING

1. The Boom tilts as the milling spindle below it moves to & fro on the lathe bed. (As drwn here, the Motor Mount (Part #5) is too close to the Suspension Hook (Part #1) to permit much tilting. In fact, the motor mount would be located further to the left than it appears in my drwg.) The split-and-clamp feature of the Suspension Hook permits one to orient the Boom as needed, and to then clamp the Suspension Hook up solid, or leave it slightly loose, so the Boom can swing as well as tilt.

2. The Boom is made from an unthreaded 4-1/2' length of 1/2" black iron pipe, which can be got at any well stocked old time hardware store, or from a company that sells & installs residential/commercial heating equipment.

3. The Hanger is shown schematically only, in the GA drwg above. See notes and Detail A in connection with Part 1, on the following pages. Above all else, the Hanger must be so made and mounted as to preclude the remotest possibility of the "Overhead Gear" ever coming adrift and crashing down on the lathe below.

4. The size and placement of the counterweight "W" will vary with the work in hand, heavy work requiring more weight than light. Try a weight of about 7 pounds initially.

Part 1
Suspension Hook

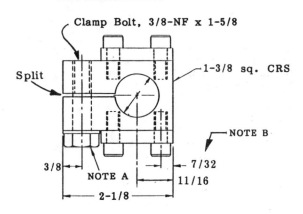

Clamp Bolt, 3/8-NF x 1-5/8

Split

1-3/8 sq. CRS

NOTE B

3/8

NOTE A

7/32

11/16

2-1/8

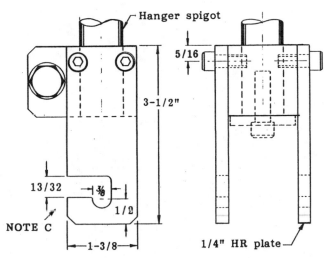

Hanger spigot

5/16

3-1/2"

13/32

3/8

1/2

NOTE C

1-3/8

1/4" HR plate

NOTE A: Not shown, but necessary, to give wrench clearance, is a washer of suitable thickness under bolt head. Alternatively, weld a permanent handle onto the bolt head, or be really fancy, and fit a clamp-on ball handle.

NOTE B: This 7/32" dimension should be closely adhered to, so that the four 1/4-28 x 5/8" soc. cap screws which secure the side plates of the Suspension Hook to the Clamping Block will not enter the 0.800" bore in the Clamping Block; these screws are NOT meant to screw in against the Hanger spigot.

NOTE C: A plain 3/8 reamed hole could be used instead of the slotted entry to the 3/8" pivot hole, but the latter permits the Boom, etc. to be easily and quickly dismounted for repair or overhaul.

For efficiency: cut 7 pieces of 1-3/8" square CRS of appropriate lengths for Parts 1 thru 5, leaving a machining allowance on each. Bring to finished length specs by facing each sawn face, one after the other. Similarly, bore the 0.840"* 'pipe hole' in Parts 2 thru 5 all at one session; make the holes an easy slide fit on the pipe. (Note that the hole in Part 1 is NOT the same size.)

* Machinery's H'book shows the O.D. of 1/2" black iron pipe as 0.840". Measure the pipe you buy, and work to that; samples I measured at a local hardware store ran +/-0.008" on the nominal 0.840" size.

Part 2
Trunnion Block

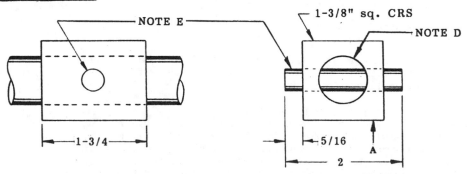

NOTE E

1-3/8" sq. CRS

NOTE D

1-3/4

5/16

A

2

NOTE D: Bore full length 1 or 2 thou over your pipe dia. A high class fit here is not required: the Trunnion Block is slid on, and then locked in place by the 3/8"Ø CRS cross pin, per Note E below.

NOTE E: After the Boom is positioned correctly in the Trunnion Block, cross drill/ream both parts 3/8"Ø together. To prevent the Pin working out, Loctite™ it in place, or fit a 10-32 setscrew per Arrow "A". Or make the Pin a little longer, and fit washers and screws (as in Detail A) at both ends. The Pin serves as the pivot for the Boom, and at the same time locks the Trunnion Block and Boom assembly together.

Part 3
Jockey Pulley Block (2 off)

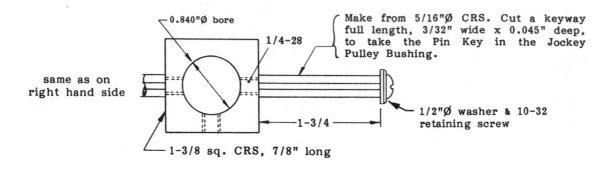

—0.840"Ø bore

1/4-28

{ Make from 5/16"Ø CRS. Cut a keyway
full length, 3/32" wide x 0.045" deep,
to take the Pin Key in the Jockey
Pulley Bushing. }

same as on
right hand side

1/2"Ø washer & 10-32
retaining screw

1-3/4

1-3/8 sq. CRS, 7/8" long

JOCKEY PULLEY DETAIL

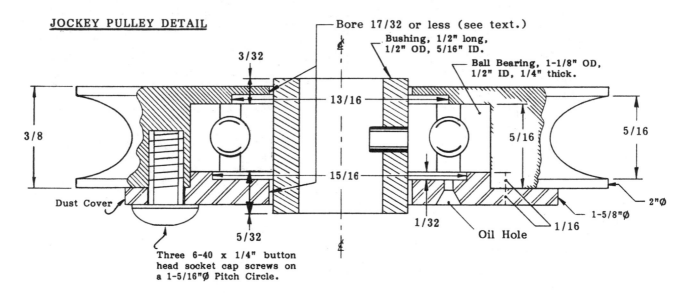

Bore 17/32 or less (see text.)

Bushing, 1/2" long,
1/2" OD, 5/16" ID.

Ball Bearing, 1-1/8" OD,
1/2" ID, 1/4" thick.

3/32

13/16

3/8

15/16

5/16

5/16

Dust Cover

5/32

1/32

Oil Hole

1/16

1-5/8"Ø

2"Ø

Three 6-40 x 1/4" button
head socket cap screws on
a 1-5/16"Ø Pitch Circle.

NOTE: Bushing is made a press fit in bore of ball race. Ball
race is a push fit in 1-1/8" Bore of Jockey Pulley. Dust cover
and Jockey Pulley do not touch Bushing.

NOTE: Fit a 3/32"Ø pin into the wall of the Bushing, so that it
protrudes about 0.040" into the bore of the Bushing. Pin can be
a drive fit; Loctite is easier. The Pin need not be exactly
centered in the length of the Bushing, but it should be smack on
the diameter - i.e. this is a careful by-the-numbers cross drilling
job.   The Bushing is intended to slide freely along the length of
the 5/16"Ø Pins in the sides of Part 3, but not to rotate on same.
The Jockey Pulley turns with minimal friction on the ball bearing,
the inner race of which, and the bushing, as noted above, do not
rotate.

FOOT NOTE: Tim Smith suggested the use of a sealed
bearing in the four jockey pulleys. This would
eliminate the need for the dust cover and the monthly
drop of oil. I suspect that when I make mine, I will
do as he suggests, but retain the dust cover.

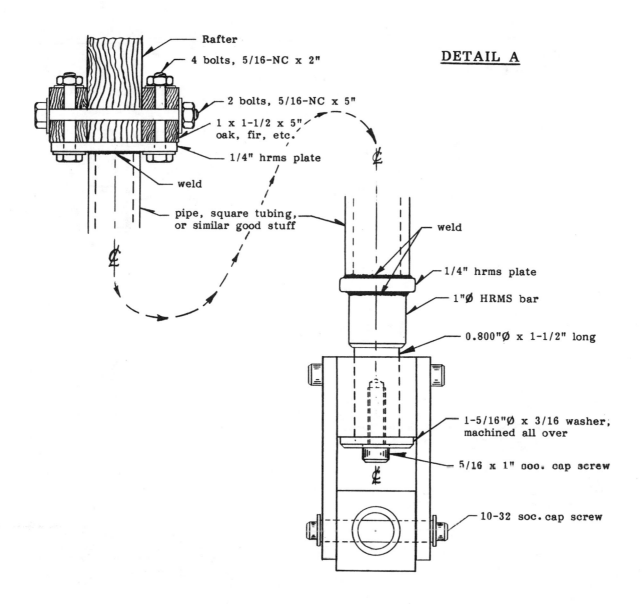

DETAIL A

Rafter

4 bolts, 5/16-NC x 2"

2 bolts, 5/16-NC x 5"

1 x 1-1/2 x 5"
oak, fir, etc.

1/4" hrms plate

weld

pipe, square tubing,
or similar good stuff

weld

1/4" hrms plate

1"Ø HRMS bar

0.800"Ø x 1-1/2" long

1-5/16"Ø x 3/16 washer,
machined all over

5/16 x 1" soc. cap screw

10-32 soc. cap screw

NOTES RE THE HANGER

Above all else, the Hanger must be so made as to preclude the "Overhead Gear" ever coming crashing down on the lathe below.

The original drwgs in ME showed the use of a piece of 1/2" black iron pipe – the same material as used for the Boom – threaded at both ends, the hole in the Suspension Hook being screwcut to match. I think the arrangement shown in Detail A above is, if not simpler, then better. It avoids both internal and external screwcutting. It can easily be made up on a lathe not large enough to pass 1/2" black iron pipe thru the spindle bore. On the other hand, it does require some welding, which will not please everyone.

Alternatively, one could make the whole length of the Hanger from solid 1"Ø hot rolled MS bar, and either weld it to the Mounting Flange at the top end, or use a screw and washer much as at the bottom end. If you go the latter route, note that the Hanger should not be free to rotate in the Mounting Flange. The screw and washer feature at the bottom of the Hanger precludes the whole rig from falling down on the lathe, ever, period. That is its whole and only job. Give the screw a shot of Loctite™ on final assembly of the Suspension Hook to the Hanger.

Part 4
Weight Hanger

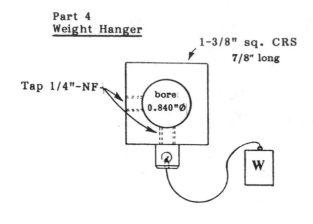

Tap 1/4"-NF

1-3/8" sq. CRS
7/8" long

bore
0.840"Ø

W

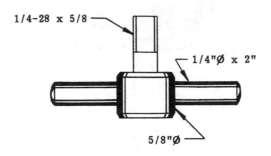

Make 4 of these T-bolts, one for each of Part 3, one for Part 4, and one for Part 5.

1/4-28 x 5/8

1/4"Ø x 2"

5/8"Ø

Part 5
Motor Mount

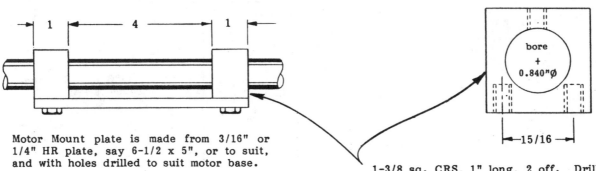

1
4
1

Motor Mount plate is made from 3/16" or 1/4" HR plate, say 6-1/2 x 5", or to suit, and with holes drilled to suit motor base.

bore
+
0.840"Ø

15/16

1-3/8 sq. CRS, 1" long, 2 off. Drill & tap three holes 1/4-28, each piece; lower two are for bolts from Motor Mount Plate, and upper one is for a hex head cap screw to lock motor at any desired location on the Motor/Pulley Boom.

MOTOR PULLEY

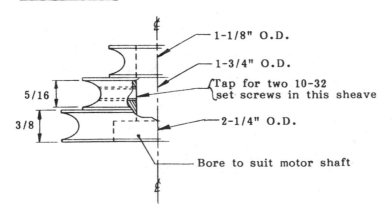

1-1/8" O.D.

1-3/4" O.D.

Tap for two 10-32 set screws in this sheave

2-1/4" O.D.

Bore to suit motor shaft

5/16

3/8

46

HOW TO MAKE A KEROSENE BURNING BLOWLAMP
by "Octogenarian"
(Reprinted with permission, and with minor editing, from an article
which appeared in *Model Engineer Magazine* for November 17, 1932, p467.

Introductory comments

What the English call a "blowlamp" is to me a "blowtorch"; in most places throughout the material which follows I have retained the term used by "Octogenarian".

You might wonder, as I did, whether there is any reason to even consider trying to make a blowlamp run on kerosene - maybe propane is just as good or better. I contacted one of the big oil companies here, and found that kerosene and propane have almost identical Btu/lb values, (20,000 vs 19,800). According to the guy I talked to, it is difficult to get a blowtorch to run on kerosene (But see below!) but he said his runs fine on Varsol. Winter diesel should be ok as an alternative to kerosene, but the problem is to get the torch hot enough to burn well on this heavier fuel - kerosene is heavier than say white gas, and therefore does not vaporize so readily. Also cost might be against kerosene. So went my discussion with the guy from the big oil company.   (Varsol is paint thinner.)

So why include something on making a blowlamp that'll run on kerosene? Well, just one month after this article appeared in *M.E.*, somebody else wrote in to the Editor, having just finished making a blowlamp to "Octogenarian's" design. The letter writer says:

" -- I find it is the easiest and finest burner I have ever made. It can be made by the average amateur simply by following Octogenarian's instructions. I made mine with a 15/16" nozzle. With 15% petrol (gas)/85% kerosene fuel mix and a 5" flame, I can melt 1/4"Ø copper rod quite easily in one minute, until it runs like white metal. When the regulator is turned down, but with the same pressure, I have silver soldered a nipple on a 1/16" copper pipe. With 25 lbs pressure from a self contained pump and (with) regulator full open, I can get an 11" flame for eight minutes. I then begin to lose heat because the blast is so fierce that it cools the vaporizing coil and the lamp pulsates at times, causing a white flare. After a few minutes, when the needle valve was closed a bit, it soon picked up again. I have left the lamp untouched for 20 minutes with flame turned low, and then opened up again and got a 5" flame with 5 lbs pressure. -- I have 4 blowlamps, and apart from my 5 pint one, this one is best. This lamp has a 1.5 pint tank, hence the long time of burning - about 2 hours during tests. -- "

With that recommendation, I feel this item has a place in this book. I hope you find it interesting.

Note: here and there where the original author made reference to a British thread, I have translated this to the nearest equivalent North American thread. GBL

HOW TO MAKE A KEROSENE BURNING BLOWLAMP
by "Octogenarian"

I have had blowlamps of the "Primus" type in more or less constant use ever since there have been any to use. I have worn out no end of burners, but do not recall ever having had a worn-out container. I have sold them, and had them stolen, but never worn one out. It is therefore not surprising that a few years ago I found myself in possession of a good container minus a burner, and was seized with the idea of trying to make one for myself. I was under the impression that it was by no means an easy job to design a blowlamp which would burn kerosene in a really practical and satisfactory manner, and that the Primus principle was practically the only one that worked really well. I had seen the results of attempts made by several other workers, but none of them could even charitably be described as successful and reliable.

They could all be summed up pretty well as follows: They could be made to burn, so long as they were pre-heated enough, blown up enough, air supply adjusted with micrometer-like accuracy, etc. But there was no certainty that they would not suddenly go out at any moment, or suddenly squirt flaming kerosene, or only burn with violent fluctuations and deafening noise. Anything in the nature of control seemed to be unknown, impossible, or both. I did not, therefore, expect to make an easy meal of the job, and I didn't!

The only kerosene burning blowlamp - other than a Primus - that I have ever handled was a huge affair holding about a gallon. It was given to me by somebody who must have held a secret grudge. No power on earth would make it burn, but as an incendiary engine it was without rival: nothing in the same room with it - including the operator - was safe for a moment from its incontinent eruptions.

Still, it was said that in the past it had burned kerosene with success, so I thought that if I produced something akin to it on a small scale I might be able to tame it.

Early Trials
Numerous experiments were made and the job became rather interesting, although at the end of some weeks "the perfect lamp" seemed to be as unattainable as ever; when suddenly, after some trifling alteration, the lamp burned fiercely and properly.

The flame was turned down by means of the adjustable nipple, and it still burned all right. The pressure was lowered, and the flame just calmed down, but did not go out or turn smoky. The pressure was then blown up to the maximum obtainable with the small self-contained pump, and the flame tested for temperature. It was found to be capable of doing brazing jobs that an ordinary pint-size Primus pattern lamp simply could not look at.

It must be confessed that Luck must have been very busy that day, because several alterations which were subsequently made, and which were confidently expected to make the thing better still, had just the opposite effect.

The obvious conclusion is that if these notes tempt anyone to try making one of these burners, they will be well advised to follow the particulars given herewith pretty closely.

It may be mentioned that this lamp - like all other kerosene-burning lamps I have ever handled - gives its best performance when 10 percent to 20 percent of petrol (car gas) is added to the fuel to liven it up. If anyone likes to try it, they will find that this lamp will burn almost as well with petrol alone.

The Three Vital Factors
I do not suggest that no-one else has ever made a successful kerosene blowlamp... I claim nothing but colossal luck as the reward for a little patience. One or two notes with regard to the principle of the thing may be of some interest.

For one thing, the size of the nipple orifice does not appear to have much to do with it, because the lamp burns just as well with the full opening - about 0.018" - as it does when the opening is greatly restricted by the point of the needle, as is the case when the flame is turned down to a minimum. It is also interesting to note that if, when the container is pumped up to full pressure, the flame is turned down to half its full size, or even considerably less, it will burn with the same fierce heat that the full flame does. No ordinary Primus pattern blowlamp will do that.

The amount of air admitted at the nipple end of the Flame Tube apparently requires no regulation at all so long as there is an unrestricted supply. It will be observed, upon reference to Fig. 1, that the area of the air holes (at the nipple end of the Flame Tube) is so great it is hardly possible more air could get in if the back half of the Flame Tube were cut off altogether. Also, the lamp burns just about the same if half these holes are masked by a tin slide slipped on to the tube.

My experiments seemed to point to the following three vital factors:

1. The size, shape, and disposition of the slots in the Flame Tube which lie under the vaporizing coils.
2. The length and position of the coil.
3. The amount of nozzling-down of the end of the Flame Tube.

(But note that the chap who wrote the glowing letter of praise which appeared a month after this article was published, had in fact nozzled his Flame Tube down just slightly more than shown by Octogenarian. GBL)

Also, when burning kerosene unmixed with petrol, it seems essential to have some sort of baffle fixed in the Flame Tube, somewhere about the center of the coil, on which the flame can play, and thus keep red hot. Several things were tried, such as a plate with a large hole in the middle, which allowed the majority of the flame to pass through, while deflecting the remainder sideways and out through the slots under the coil. Nothing, however, was found to be quite so effective all round as the small piece of iron rod - a nail, in fact - which passes through between the coils and across the Flame Tube. (See Fig. 1.) This of course will burn away in the course of a few months' hard work, but being merely part of a large wire nail, is not an expensive item.

The drawings pretty well explain the whole thing, but a few hints may be useful before the reader proceeds with construction.

Fig. 1 - General Arrangement Drawing

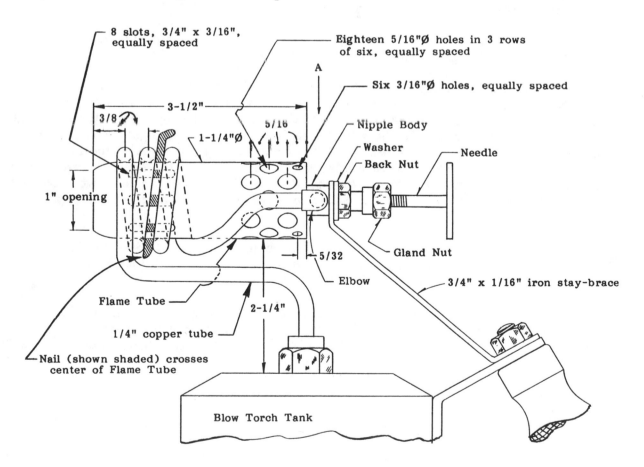

Choice of Materials

When making the Flame Tube, use good stout tube. It gets tremendously hot at the nozzle end, and if made of less than 18 gage*, as directed on plan, it would be very prone to melt if a specially strenuous brazing job was in hand.

> *Octogenarian does not say to which gage he is referring, but given that he is speaking of sheet copper, I think it is likely that he means "Birmingham Gage for Sheets"; if so, this would be about 50 thou wall thickness, while the 10 gage copper pipe specified for the coil would be 0.125" wall thickness. GBL

As a matter of fact I did once, in those very circumstances, cause the end of the Flame Tube to bulge and distort to such an extent that I had to make a new one. I therefore made a steel one. While it worked just the same and could not be melted, it was rather troublesome to make, and soon scaled all away at the end, so another brass one was made.

The copper coil should also be made of heavy gage stuff, the heavier the better. I have just had to replace a 10 gage coil which has lasted only about 18 months. Still, that lamp has probably generated as much heat in that time as the majority of Primus lamps generate in ten years, so it is not too bad.

Making the Flame Tube

A good way to nozzle down the end of the Flame Tube, if the operator is not practiced in that sort of work, is as follows:

First cut off and square up the ends of the tube; cut the tube an inch or so too long at first, so that if the nozzling down is carried too far, or otherwise spoiled, some can be cut off and the thing corrected without wasting the whole tube.

Now turn a 1/4" radius on the end of a foot or so of 3/4" iron rod. Fix this horizontally in the bench vice, with a trifle more than the length of the Flame Tube projecting.

Thoroughly anneal one end of the tube, and slip it over the rod so that the annealed end of the tube is about the middle of the radius. Hold the tube down tightly on the rod, and with a small wooden mallet tap the extreme edge of the tube down towards the rod. Turn the tube slightly and tap again. Keep this up until the whole of the end of the tube is slightly and evenly curved inwards. Re-anneal the end and repeat the tapping and turning process with the tube advanced 1/8" further along the rod. If the tube is kept well annealed, and the operator carefully notes the effect of each tap, he will soon get the knack of the thing, and be able to work the end of the tube down to the required size and shape in a few minutes, whereafter it is cut to final length.

The thickness of the End Plate or Disc (Fig. 7) is unimportant, but it must be fixed either by silver soldering, or by turning it a driving fit in the Flame Tube, driving it in, and slightly burring the end of the Tube over it to hold it.

Drilling the several air holes in the Flame Tube, as shown, is rather a tiresome job, and I am quite sure it would answer just as well if a couple of large square openings were cut in the same way as is found on the standard Primus flame tube. Do not, however, drill or cut away the End Plate in any way. I have found drilling holes in the back plate of an ordinary steamboat pattern blowlamp tube sometimes has a surprising effect - and not always a good one - on the performance of the lamp. Doing so might similarly upset the working of this one.

Take care to set out the eight slots that lie under the coil exactly to the size and position shown. These can, of course, be drilled and filed out, but it is much better and quicker in the end to set up the drilling spindle on the top slide, at center height, drive it with the "overhead", and cut the slots with a small slot drill, the tube being held in the chuck, and indexed the necessary distance between the cuts. However, anyone having the facilities indicated will not need to be reminded of their capabilities.

The Vaporizing Coil/Fuel Tube

If the 1/4" copper tube for the Vaporizing Coil is well annealed it can easily be bent to shape around a 1-1/4" bar held upright in the vise. Great care, however, must be taken to form the coil, and finally fix it, so that the coils are exactly in the position shown on the drawings.

The Fuel Tube is fixed to the blowlamp fuel tank by bushing the hole in the union, after sawing off the old burner, and silver soldering the tube into the bush. (Take the bush out of the container first!! GBL)

The Nipple and Needle Valve Assembly

The details of the Nipple Body (Fig. 2) are slightly different from those of the one described in my earlier article describing "A Blow-lamp Improvement" in the June 30, 1932, issue of *M.E.*, but the process of making it is exactly the same, and the drawings are largely self-explanatory. If the reader has not noticed the previous article, it may be worth while to refer to it for particulars of the method of drilling the 0.018" nipple hole. (*Note: Pertinent material from that article is given at the end of this section. GBL*)

Fig. 2
Details of Adjustable Nipple
for a Kerosene Burning Blowtorch

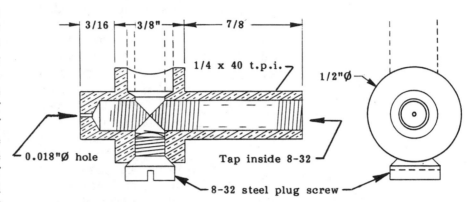

NOTE: The 7/8" long x 1/4"Ø (or 5/16"Ø - see text) portion of the Nipple Body is screwcut 40 (or 32) t.p.i. for its full length. The 3/16" x 1/4"Ø spigot at the front end of the Nipple Body is just plain 1/4"Ø: this spigot sticks through the 1/4" hole in the center of the disc at the back end of the Flame Tube.

The Elbow (see Fig. 5) on the end of the vaporizer tube can be sawn out of a small block of copper. Two 1/4" holes are drilled in it at right angles until they meet, and the outside filed to shape*. In making and fixing this Elbow, pains must be taken to get a good clear passage through it, as this is a point at which scale is apt to collect and block the pipe.

> *NOTE: This humble machinist, who has another 40 years or so to go before he could fairly adopt the pen name "Octogenarian," would like to observe that there is, in his opinion, no sound reason to file the outside of this Elbow to round contours. I didn't notice this while making the GA drawing, and drew it as Octogenarian showed it in his. However, I show it made from a simple piece of rectangular section material in the scrap view. This change makes the Elbow simpler, faster, and easier to make.

> Note too that Octogenarian says to "take pains" re the drilled passage. I would say, yes, put your effort and time in there, not into fiddling about with a file to produce artistic but completely meaningless contours on the outside. GBL

The Plug inserted in the "clearing hole" should be coned on the underside of the head, as shown, and should be made to bed securely into the slightly countersunk hole. Use a little bit of fine abrasive on the cone only, not on the threaded portion, and work the two parts together until a nice seat is formed.

> NOTE: If the maker elects to dodge this little bit of work, and simply close the hole with a flat-headed screw, trusting to packing under the head to keep it tight, he will saddle himself with permanent trouble. I have never succeeded in keeping such an arrangement gas-tight for any useful period.

This Plug is not absolutely essential, but greatly facilitates the cleaning out of the burner. A lamp of this pattern seldom or never suddenly chokes, after the exasperating fashion of the Primus variety, but after considerable use its action will abate somewhat. In that event, if the Needle is withdrawn, the clearing plug removed, and the coil tapped and shaken, a quantity of dusty deposit can be shaken out. This will be assisted if the burner is detached from the container and energetically blown through during the tapping and shaking business.

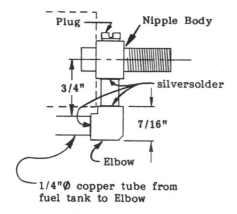

Fig. 5
Plan view (view on A, GA drwg)
of Nipple Body

Plug

Nipple Body

silversolder

3/4"

7/16"

Elbow

1/4"Ø copper tube from
fuel tank to Elbow

As the 1/4" copper pipe is not nearly stiff enough to carry the Burner Assembly (i.e. the Flame Tube plus the Nipple Body) without other support, a Stay, or Brace, must be added, as shown in Fig. 1, but as the handle parts of blowlamp tanks differ considerably, the maker may have to devise some special way for anchoring the Stay at its bottom end. However it is done, keep the burner perched up at a good height above the fuel tank, as the lamp is more pleasant to handle if the tank keeps fairly cool. It will be found, however, that a tank fitted to one of these special burners does not get nearly so hot as one fitted with its usual burner.

It may occur to someone to make one of these burners, and a tank for it as well. If you do so, and elect to adopt the simple expedient of raising the tank pressure with a bicycle pump, you must do so with caution: it must be realized that the very short pump fitted in the Primus container is incapable of raising the pressure above a very low value, no matter how many strokes are given. Investigating this matter with a somewhat dubious Dunlop tire pressure gauge, I found I could never get a pressure exceeding 25 lb. per sq. inch in the container. On the other hand, a few strokes of a long thin bicycle tire pump may easily raise the pressure to 60 to 70 lb., and it is more than likely that the burner would have to be drastically altered to make it burn at that pressure.

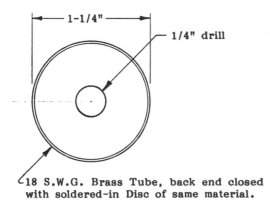

Fig. 7
Rear view of Flame tube

1-1/4"

1/4" drill

18 S.W.G. Brass Tube, back end closed
with soldered-in Disc of same material.

WHAT CAN IT DO?

A 4" length of 1-3/4" OD steam tubing, which is pretty stout stuff, had a 1/4" hole drilled in the side. The end of a 2" length of 3/4"Ø steel bar was turned down to form a 1/4"Ø tit on the end, and this tit was driven into the hole. This was not packed tightly round with coke, to aid the heating process, but just laid in a slight depression on a bed of coke. Starting all cold, the joint was perfectly brazed with brass wire in eight minutes. (Try that with a propane torch! GBL)

(Note: Although the drawings, which continue on the following pages, may appear to be out of sequence, they are not: these parts are dealt with in the text which also follows.)

WARNING: The construction and use of a device to burn white gas, car gas, kerosene, or anything else of a like nature, is not something to be undertaken without paying close attention to what one is about, and seeing to it that a first class job is made of the matter from start to finish. If you have any doubts as to the safety of the finished product, test it first with a very small amount of fuel in the tank, and then give it a longer test while you stand well away. Have a fire extinguisher handy at all times, both when testing, and later when using the lamp on a routine basis, and pay full attention to all other standard and common sense fire and safety precautions, INCLUDING NOT STORING OR USING SUCH EQUIPMENT IN AN UNVENTILATED AREA, OR BETTER STILL, NOT USING IT INDOORS AT ALL. The alternative might be that you (or worse, someone else) might be forced to spend a few months in the burn unit of a hospital.

GBL inserts some **Notes re Making the Nipple Body**

One of the things not made entirely clear in Octogenarian's text and drawings is that the Nipple Body (Fig. 2) is (theoretically) made from 1/2"Ø brass bar, with the ends turned down to 1/4"Ø for distances of 7/8" and 3/16" respectively. I say "theoretically," because I would be inclined to make this part from 9/16" or 5/8"Ø brass bar, as follows:

I would start with a piece of say 5/8"Ø material about 1-7/8" long, i.e. somewhat longer than the finished OAL of 1-7/16". I would mentally put the extra material at the nozzle hole end, and I'd stick it in a 4-jaw chuck, indicate it to run true, and then drill and tap the deep 8-32 hole. In doing this, I'd use a D-bit for part of the tap drilling operation, to ensure that the hole did not wander; it MUST be straight, because of the very fine point on the end of the Needle (Fig. 4, below) which screws into this hole.

Next, with the Nipple Body still undisturbed in the chuck, into said 8-32 tapped hole I would screw an 8-32 threaded plug, previously made up and ready to use. Once I had that plug screwed in solid, I would put a small center hole in the exposed end of said plug, de-chuck, and reverse the Nipple Body in the chuck, indicate as before, and center drill the opposite (nozzle) end.

I would then set up the job between centers and turn the OD to 1/2, 1/4, and 5/16 or 3/8"Ø (see **"Put enough "meat" in the Nipple Body"**, below) and screwcut the outside of the 7/8" long portion to 32tpi for about 3/4". This done, I'd stick my special shop-made 0.018"Ø drill in the lathe headstock, and slide the Nipple Body onto it, carefully finger feeding same on and off the drill until the hole was done (just as Octogenarian tells us to do where he speaks (below) of refurbishing this hole in the future).

Cross drilling the Nipple Body for the 1/4"Ø vaporizer coil and the Clean-out Screw is routine. The countersink on the clean-out screw is drawn with a 90° included angle, but could be 82°, which is the standard countersink angle in North American practice.

> (It is an easy matter to make up a special 90° (or 82°) countersink cutter of D-bit form, but with the flat taken in far enough to remove slightly more that half the diameter, rather than the opposite, as is normally done when making a D-bit. (See **TMBR**, page 52.) Made as here described, such a countersink will not chatter, but may exhibit a tendency to throw in a burr unless kept sharp by stoning the face of the flat. GBL)

Put enough "meat" in the Nipple Body

Octogenarian called for both the long and short ends of the Nipple Body to be made 1/4" O.D., and on the section drawing of the Nipple Body (i.e. at Fig. 2) this dimension is used. However, if you put an 8-32 thread into a piece of 1/4"Ø material, the wall thickness is reduced to 0.043". If you then cut a 32tpi thread on the OD, the effective wall thickness drops to 0.0159". (A 40tpi thread will bring this up to 0.0214".) Considering the duty involved, (conducting kerosene to the point of burning) and the fact that the Nipple Body serves in part as a structural member of the completed blowlamp, I feel that 1/4"Ø for the back end of the Nipple Body is FAR TOO THIN.

From preliminary calculations, it looks like a torque of less than 3 ft.lbs. would cause the Nipple Body, which is effectively a hollow brass bolt, to fail; this assuming the use of brass having a yield strength of 50,000 psi.

I would say that 5/16"Ø would be the minimum to consider, for the long end of the Nipple Body, and even 3/8" would do no harm. In the General Arrangement drawing (Fig. 1) the after section of the Nipple Body is drawn with a 5/16" OD. This will leave you with a wall of 0.0526" - i.e. nearly 2.5 times as thick - which would be significantly stronger.

I would most strongly recommend you make this change, which will of course also require changes in the dimensions of the Back Nut (Fig. 1) and the Gland Nut (Fig. 3).

Some notes on drilling the Nipple Hole, and related matters

(excerpted from Octogenarian's earlier article in *M.E.* of June 30, 1932, p. 613)

...the nipple hole....should not be more than 0.020"Ø. The exact size is unimportant, but if under 0.018", it may not be possible to get quite so large a flame as the blowlamp ought to be capable of producing, and if over 0.02", it will probably be found necessary to always keep the lamp more or less throttled down when working at full pressure.

...There is no easy* way of drilling this hole, altho no doubt anyone who has dabbled in watch or clock work would regard it as a veritable tunnel. It must be observed that it has to be drilled as an extension of the long hole in which the Needle lies, and that it must be exactly central to said hole.

> *(And if I might intrude for a moment, to add a word of encouragement, consider the tools at your disposal today: at the very least you have a powered lathe, and probably one much better than what "Octogenarian" had to work with 50+ years ago - his lathe was quite likely powered by a foot treadle. Furthermore, his pen name implies that he was pushing 80-odd years, so don't be put off, thinking, "Oh, I could never do that!", GBL)

Even if a drill of the required size were available long enough to reach down the long hole, it would have to be driven at a speed far in excess of the speeds normally available on the average amateur's lathe, in order to make it cut without springing, bending, and probably breaking. Also it is highly unlikely that it would find the center with the necessary certainty.

Of course, the Nipple Body could be finish turned and parted off without boring the small hole; the job could then be reversed in the chuck, and the small hole drilled with a very short drill held in a pin chuck. This method, however, would require great accuracy in the re-chucking.

I have made many of these Nipples for flash steam boilers, and have found the following method of drilling the nipple hole well repays the trouble taken, as it ensures the absolute accuracy which is vital for the success of the lamp.

The first step is to make a special drill for the job. Procure a piece of drill rod about 3" long and of exactly the core diameter of the screwed hole in which the Needle is to work. (Buy a length of drill rod of the appropriate size. GBL) This piece of drill rod is chucked, and indicated to run dead true, with about 1/4" protruding beyond the chuck.

The end is then turned down to 0.018" for a length of exactly 1/16", to form the drill itself, and the end of the rod at the root of the drill beveled off to the angle of an ordinary drill point. The turning tool used to make this drill must have a very fine point with plenty of top rake, must be oil stoned to absolute sharpness, and must be set precisely at center height. Before starting the job, set the top slide over about 1°, to ensure the drill itself comes out a shade larger at the tip than at the root.

Admittedly, the turning of this drill is a delicate matter, but if the operator keeps watching the job through a magnifying glass after it has got below say 1/32" diameter, he will soon get control, and see how much cut it is safe to put on, and how fast to feed. The thing which really requires the greatest care of all is to avoid bending the drill when applying the micrometer.

The turned point can then be flattened on opposite sides, and the point brought to the form of a spear point drill by cautiously rubbing on an oil stone, with the aid of a magnifying glass. The drill is then hardened and tempered dark straw in a very small gas flame, being careful to keep the actual point just clear of the edge of the flame, otherwise it may be burnt.

As stated above, the drill must be exactly 1/16" long, for the following reason: I have found from long experience that if the nipple hole is 1/16" long, it will be right. If it is longer, the Needle point will have to be too long and thin to be reasonably durable; if it is shorter, the tendency of the Needle to gradually enlarge and distort the hole is much increased.

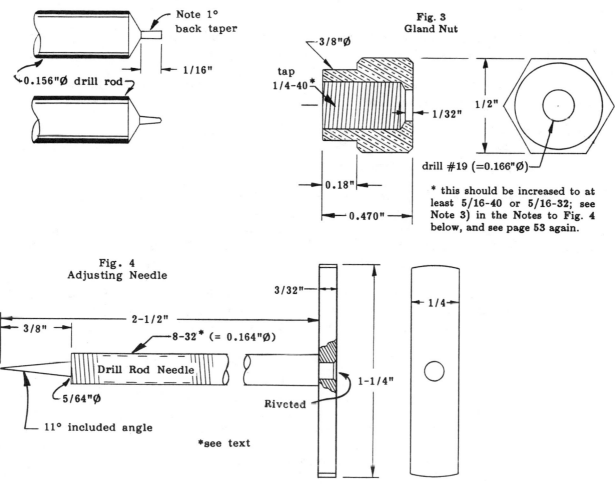

Fig. 6
Details of the special drill which must be made to drill the 0.018" hole in the end of the Nipple Body.

Note 1° back taper

1/16"

0.156"Ø drill rod

Fig. 3
Gland Nut

3/8"Ø

tap 1/4-40*

1/32"

1/2"

drill #19 (=0.166"Ø)

0.18"

0.470"

* this should be increased to at least 5/16-40 or 5/16-32; see Note 3) in the Notes to Fig. 4 below, and see page 53 again.

Fig. 4
Adjusting Needle

3/32"

1/4

3/8"

2-1/2"

8-32* (= 0.164"Ø)

Drill Rod Needle

5/64"Ø

1-1/4"

Riveted

11° included angle

*see text

NOTES

1) Make Gland (Fig. 3) and Nipple Body (Fig. 2) from brass or stainless steel.

2) Make Clean Out Screw (Fig. 2) from drill rod; harden, and temper to a blue color.

3) You may choose to make (or buy) one or more special taps, e.g. 1/4-32, (or, better by far, 5/16-32, or 3/8-32 - see text) to match the pitch of the 8-32 thread on the Needle. These taps would be used to tap the Gland Nut and the Back Nut. (The latter item is not shown - see Note 4 immediately below.) Corresponding male threads would naturally be screwcut, obviating the need to make or buy a die.

4) The Back Nut (not shown on this drwg: see Fig. 1, the GA drwg) is made from 1/2"A/F brass hex material. Make it about 3/8" thick, and tap to match the thread on the Nipple Body.

5) The Needle (Fig. 4) is not drawn exactly to scale - go by dimensions given. Other parts are to scale.

In making the Nipple Body (as detailed in the next paragraph), finish turn and part off slightly longer than necessary, to ensure the small hole is not touched by the parting tool. Then reverse the Nipple Body in the chuck and carefully face off the end with very fine cuts until the hole appears. As the small drill was previously run in up to its shoulder, it is obvious the finished hole must be exactly 1/16" long.

The best thing to do after chucking the brass stock for the Nipple Body is to first drill and tap the hole for the Needle to the full depth, then put the special shop-made 0.018" drill in a small chuck - hand held - and with the lathe running at top speed, pass it into the hole and press it in gently for a moment, draw it back slightly, and then press it in again. Keep thus gently poking it in, and in a few seconds it will be found to have gone in right up to its shoulder. If you try to feed it right in, all at one hit, it is almost sure to break.

Do this drilling as the first operation, as it is the only part likely to cause trouble. Once this hole is done, all else is a certainty.

When turning the point on the Needle, proceed similarly as with making the drill, so the point runs dead true with the shank.

Make the Gland at least as long as shown*: the more packing that can be got in, the tighter the joint will be, and the longer it will last.

> *(i.e. 9/16" long; for some reason, Octogenarian shows a Gland Nut only 3/8" long in the Nov.'32 article. In my drawings, I have shown the Gland Nut as 0.47" long - a compromise between Octogenarian's two dimensions! GBL)

It should be noted that the Needle does not have a parallel part working in the Gland, as is usual in the glands of ordinary valves, but has the threaded part working in the packing itself. I have proved to my own satisfaction that this arrangement is infinitely the more effective and durable; it lasts almost indefinitely, and stays gas-tight while letting the Needle work much more freely than is possible with the usual arrangement.

The Gland must be packed with asbestos* string saturated with a paste of graphite and paraffin, as much as can possibly be got in. It will probably leak at first, but after being tightened up two or three times, when hot, will become permanently tight.

> *See note re asbestos and alternatives thereto, below. GBL

There is a minor point in the construction of adjustable nipples worth noting. It is desirable, if possible, to make the screw threads on Needle and Gland to the same pitch. If this is done, the Needle and Gland can be removed from the Body together for cleaning, without withdrawing the Needle from the Gland, and so, possibly, disturbing the packing. This cannot be done if the threads differ in pitch; if they do, then the Needle must be entirely withdrawn before the Gland is moved. If the two are forced off together the packing will be broken up and will certainly leak when the parts are re-assembled.

A convenient combination for a small nipple is to screw the body 7/32-40, and fit a needle having a 1/8 Whitworth thread, which is also 40 t.p.i.

Unfortunately, in making the Nipple (discussed, but not illustrated in this book.GBL), it is not easy to abide by this rule, as... a Needle less than 5/32" diameter (is considered by Octogenarian to be) not stout enough for a lamp in constant - and more or less rough - shop use....

> Note: Octogenarian's drawings for the June 30, 1932 article (from which this excerpt is taken), and those which accompanied the later Nov 17/32 article on "Making a Kerosene Blowlamp", show a 5/32" Whitworth thread (= 32 t.p.i.) on the Needle, and a 1/4-40 thread on the Nipple Body/Gland Nut. This is easy enough to correct if you want to do the job right: make or buy suitable taps, and screwcut the mating threads to match - which is exactly what Octogenarian then went on to say he planned to do in making another such Nipple/Needle/Gland himself. See my notes re this on my drawings. GBL

In use, the Adjustable Nipple should not be used as a valve to turn the lamp out when shutting off after use, but only to adjust the size of the flame.

> NOTE: in the design given here for a kerosene blowlamp by "Octogenarian", the Adjustable Nipple is the only valve on the burner, and would therefore have to be used for shut off as well as flame control. GBL

The only exception to this rule is as follows: After considerable use it may be found upon lighting the lamp that the flame is not so large or fierce as usual. This is sure to be caused by the nipple hole having become slightly furred up. When this occurs, do not put the lamp out, but screw the Needle in until the lamp goes out, which it will not do until the Needle is in nearly as far as it will go. Then, very gently turn the Needle in until it will not go any further, at which point it will be found that the tip of the Needle is just peeping out of the hole. If the Needle is then quickly drawn back, the lamp can be instantly relit, and will again burn as well as ever. I have found that it is not necessary to dis-assemble the parts for cleaning out more often than about twice a year, when the lamp is in use several times per week.

I have also found that drill rod is the only satisfactory material for the Needle. Any other material tried has failed to retain its sharp point for any useful period, whereas the drill rod point lasts almost indefinitely. In turning the Needle, the right taper will be produced if the top slide is set over about 4°, (5.5° in the case of the Needle tip shown in the main article, and in my drawings herein. GBL) and after turning, the point should be brought to a fine pitch of sharpness and polish by being rubbed with an oilstone slip while running the lathe as fast as possible.

The plain steel cross handle shown is far better in use than any form of wheel or knob.

(...some irrelevant material is then left out of the material excerpted from the June 1932 article, which then continues as below. GBL)

If, after long usage, it is found that the nipple hole has become so large and distorted as to seriously affect the burning of the lamp, it can be put right in the following simple manner:

The nipple hole is drilled out, tapped 4-40, and a matching thread cut (with a die) for a distance of a little more than 1/16" on a suitable piece of brass rod; this is then screwed into the hole as tightly as possible, and cut off. The tapping size drill originally used to drill the long hole in the Nipple Body hole is then passed down into the Nipple Body and given a few turns to make sure that all is clear at the end of the hole, and that the end of the threaded brass plug is not projecting inside. The special nipple hole drill (see Fig. 6) is then fixed in a true running chuck in the lathe headstock, the Nipple Body held in the fingers and passed over it until the drill point reaches the bottom, and then gently and intermittently pushed forward until the drill point comes through.

(End of excerpt re drilling of
the fine hole, and related advice.)

WARNING/NOTE re Asbestos:

In the above article, "Octogenarian" speaks quite casually about using asbestos as a packing material. It is now well known that asbestos is a severe cancer hazard, particularly for people who also smoke. I had a call one night from Ken Allen of Coleman, Alberta, who is a chief steam engineer, and in the course of talking to him, I asked if he could tell me of a suitable substitute packing material. He suggested "Ceramic Fiber Rope", which is a refractory insulating material. Look in the Yellow Pages under "insulation materials" and with a few phone calls you should soon find somebody who can help you out. This material, in 1/2" rope size, costs about 30 cents per foot; a 12" piece would probably furnish enough packing for a hundred kerosene blowlamps. (Shortly thereafter, Ken very kindly sent me a 30" long sample of this stuff - so if you are really stuck for a source, send me $1 and a self addressed envelope - and I'll send you a piece. [And if you do this - please, oh please! - tell me what you want - don't just send a buck naked buck.])

A VISIT WITH A RETIRED GAGE MAKER

Spot Grinding & Lapping for
Flatness & Parallelism to 0.000,01"

One of the interesting aspects of meeting other home shop machinists is the opportunity to learn new techniques, etc., which can be applied in one's own shop. At one of the monthly meetings of the B.C. Society of Model Engineers, I fell into conversation with a very quiet older gentleman, whom I will call "Jake". Jake turned out to be a retired machinist, toolmaker and gage maker. He'd made Gage Blocks (Jo' Blocks) in England during WWII. He thinks in terms of 1/100,000 of an inch.

Jake offered to show me his surface grinding equipment. I took him up on his invitation one day, and saw his whole shop. I found it most interesting, and from the feedback I've had from some of you fellas, I think you might like to know more about the techniques and equipment involved in this kind of work.

Jake showed me an 8" straight edge he'd made years before. It was lapped to such a degree of straightness that when he put an "optical flat" against the back of it to demonstrate the application of the latter item, a refraction pattern appeared which he said indicated a difference of about 1/100,000 of an inch in height between two points about 2" apart. You can be pretty sure the main working edge was equally true.

From this, and numerous other tools he'd made, and showed me that day, I'd have to say Jake knows what he's doing. (I say this because some of his equipment, which I will describe below, may sound rather "mickey mouse" at first reading. But what he has, you could duplicate, and if you did, you would have the means of doing the class of work he does, which is probably much better than most of what you or I normally do, or even dream about.)

Jake visited my shop a while later, just about the time I was finishing up work on my Toolmaker's Block, which see elsewhere herein.

I had faced the rough casting to size in my lathe, and had drilled and tapped numerous holes in it for clamping purposes, plus clamping slots, etc. He said that if I was interested, he would help me lap it flat, square, and parallel within a hundred thousandth of an inch (=0.000,01"). Having already decided to have it surface ground all over at a local tool & die shop, I went ahead with that plan, at a cost of $40. I could have spared myself the expense, as you will soon see.

When my Toolmaker's Block came back from its $40 visit to the tool & die shop, I gave Jake a call and set up a date to see what he could do to improve it further. On the agreed day, I arrived on Jake's doorstep about 9 A.M., and it wasn't long before we were busy on my Block.

> NOTE: The lapping operations described here are, for the most part, probably superfluous. According to reading I have since done, lapping is only needed when making gages for measuring the very finest of tolerances, or where an exceptionally high finish is wanted, e.g. when the surface needs to be good enough to wring to a gage block. However, even if you don't want to equip yourself for lapping, don't dismiss this section of **TMBR#2** as useless - you are going to learn, among other things, how to make a low cost but highly effective surface grinder.

The first step was to lap one of the big faces flat, and we did that. **How** we did it will be described later. The **reason** we lapped it was to make sure the Block was quite flat on at least one face to start with. Read on...

With that one face flat, we next checked the Block for parallelism on the opposite face, using a

0.000,1" Dial Indicator and a large cast iron surface plate. My Block was out by a few "tenths". We would correct that first, Jake said.

NOTE: ANYONE ATTEMPTING THIS TYPE OF WORK
DOES SO AT HIS OWN RISK!!

Before I describe Jake's surface grinding arrangements, I want to say this: the equipment and the procedure by which it is used is extremely dangerous for anyone who does not possess a clear understanding of what he is doing and where the dangers lie. Anyone using this type of equipment must be fully alert at all times that he does not pass the work under the grinder in the wrong direction or otherwise mess up. If you pass the work under the wheel the "hungry" way, you are likely to get a bellyfull of your work, literally.

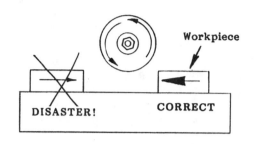

Feed the work **against** the wheel!

Let's take a look at Jake's surface grinder.
Jake's little home built surface grinder occupied about as much space as an electric typewriter. It had neither magnetic chuck, nor table feeds - nor did it need them. It consisted merely of a flat cast iron surface plate with a screw operated vertical slide in one corner, the latter carrying a grinding head driven by an electrical motor set up behind the surface plate.

You may know of the Potts Universal Milling Attachment (UMA). This item was described in *Model Engineer Magazine* in Feb. 11, 1937, page 131. The drawings which follow show a General Arrangement in pretty fair detail, of a unit much like the Potts UMA. The vertical slide and grinding head on Jake's surface grinder is very similar to the Potts UMA. Jake's equipment is of his own making, from castings off his own patterns, and the grinding wheel mounting system - see below - takes nothing from the Potts unit.

While we are on the matter of the Potts UMA, even tho' it is off the track of surface grinding and lapping, it might be worthwhile pointing out that the Potts design is intended to be used as a lathe milling attachment and can readily be adapted to serve as a lathe dividing head as well.

Also, the innards of the spindle, which are not shown in much detail in my drawings, could be made as shown in the section drawing of a Milling and Drilling Spindle earlier in this book. Or you could fit a set of *pre-loaded angular contact ball bearings*. Done right, that's about the best type of bearings there is, particularly for a grinding spindle.

How might you or I go about rigging up a simple surface grinder like this for ourselves? Buy a top quality granite surface plate and have a mounting hole put in one corner of it. Most surface plate makers will do such "extras" on their surface plates on a custom basis. One hole would add maybe $20 to the cost of a Starrett 12 x 12 x 4" Master Pink AA grade surface plate, which would be about as good a start as one could want to have.

I have a notion that somewhere there will be a reader who by now is thinkin' "Buy a granite surface plate? Shucks, man, that's gonna cost me $200!" If you prefer, you can always buy a Taiwanese factory-made surface grinder. It'll have all hand feeds, cost you 7 to 10 times as much, and will not have a great deal more capability than Jake's rig.

A plate of lesser accuracy grade would probably do for most of us, but I think the greater hardness provided by the quartz content of Starrett's "Crystal Pink" and "Master Pink" granite products might be worth the cost difference over other types of granite, particularly if one planned to do a lot of this sort of work.

The rise and fall motion, and the grinding head, are simple, altho this will take some work to produce: either make something similar to, or order drawings and castings for, the above mentioned Potts Universal Milling Attachment, available from Woking Precision Models Co. Ltd., and possibly from Power Model Supply in the USA. (see Appendix). Alternatively, a commercially made screw controlled slide mechanism could be purchased; that would add to the cost, but cut down on the time input.

Jake's motor is an old 1750 rpm totally enclosed 1/2 hp unit bolted to a little wooden board. A couple of wooden wedges are pushed in between the motor board and the base of his "surface grinder" to tighten the drive belt.

> NOTE: The following is an important safety point: It is desirable that the belt not be too tight. In fact, the belt should be loose enough so you can stall the grinding wheel on the work without stalling the motor.

I realize that this all sounds like a pretty mickey mouse arrangement, but remember, this whole thing sits on a corner of a cupboard in a crowded, cluttered basement. A better arrangement would be easy enough to devise. But even if you do set up something a little more "professional looking," don't kid yourself that you will immediately out-do what Jake does.

Now, we've had a look at Jake's little surface grinder. Next, let's see what he did to my Toolmaker's Block with it.

Jake had a variety of grinding wheels mounted on quick change arbors featuring a male taper cone which fitted a female socket in the grinding spindle.

He mounted a white aluminum oxide wheel, about 5"Ø x 1/4" wide, and dressed it by passing a diamond point under it, sliding the latter on the cast iron base plate. He then raised the wheel head, put my Block on the cast iron base plate, and fed the wheel down till it just kissed the Block.

KNOW THIS, AND KNOW IT WELL: 0.001" IS
A HEAVY CUT ON A GRINDER OF THIS TYPE!!

> Now comes the first interesting part. But keep in mind this is EXTREMELY DANGEROUS, in that we are talking about hand-held workpieces. (There should have been a guard on Jake's grinder, and there is no reason not to have one. My friend "Joe" (see p.116 in TMBR) has had wheels fly apart while he was using them, and the only thing that saved him from a hideous stomach wound, if not a slow and painful death, was the fact that he was wearing a heavy leather apron - pieces of the stone hit him in the stomach and knocked the wind out of him for several minutes. A piece of the same stone could just as easily have come up and sunk into his forehead, or whatever.

Working entirely "freehand", Jake passed my Toolmaker's Block under the wheel repeatedly, sliding it across the flat cast iron plate, always feeding against the direction of the wheel's rotation. At each pass the wheel took a light cut across the Block.

These first passes were all approximately parallel to the long axis of the Block. No attempt was made to make the cuts perfectly continuous one after the other - if he missed a patch 1/16" wide and an inch long, he didn't worry about it.

When he'd covered the surface with passes paralleling the long axis of the Block, he turned the Block 90° horizontally - thus keeping the same face down on the cast iron plate - and made another series of passes under the wheel, these parallel with the Block's short axis.

He next made a series of passes parallel with each diagonal of the face we were working on, such that by the time he was done he had passed the Block under the wheel in four different

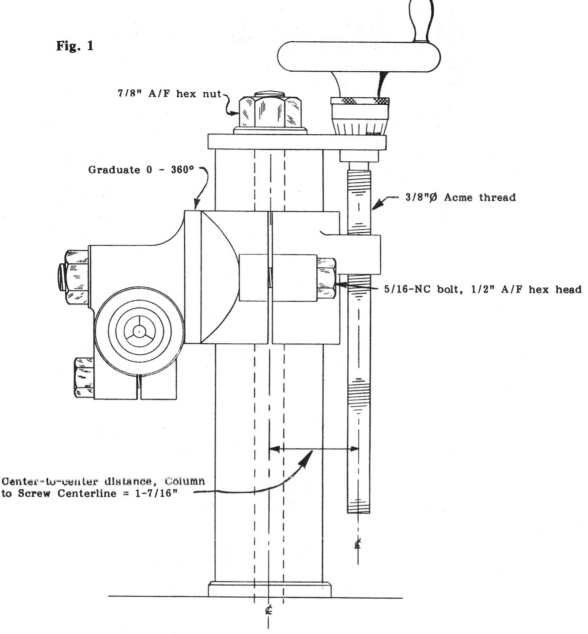

Fig. 1

7/8" A/F hex nut

Graduate 0 - 360°

3/8"Ø Acme thread

5/16-NC bolt, 1/2" A/F hex head

Center-to-center distance, Column
to Screw Centerline = 1-7/16"

A Vertical Slide & High-speed Spindle Outfit
which could be adapted for a shop-built spot grinder

Above is a General Arrangement drwg of a Vertical Slide & High-speed Spindle much like the well-known Potts Milling Attachment. Details of the spindle housing are shown on next page. The spindle housing need not be able to swivel in the vertical plane if made solely for grinding head duty. (A device of this type can also be rigged for use as a Dividing head on your lathe, by fitting a division plate to the rotating spindle and providing a detent carrier. By substituting a machine vise or other means of workholding in place of the spindle housing, it can also be adapted to use as a vertical slide for your lathe.)

This drwg was made as an accurate scale drwg: scale to obtain dimensions not given. The column is 1.75"Ø, hollow - say 3/16" wall thickness - and is drwn here slightly longer than in the original Potts design.

WARNING: If tightening the central bolt to anchor the unit to a machine table - most likely the T-slotted cross slide of your lathe - remember that over-tightening can damage the T-slot: the Column, being hollow, offers no support against tearing out the T-slot lips. Be Careful!!

NOTE: Although the use of a conventional rectangular gib key is shown, this would be a classic spot to employ a triangular gib key (see section immediately following this one), except that in this particular application, where maintenance of radial alignment is unimportant when the head is raised or lowered, the rectangular gib will serve just as well.

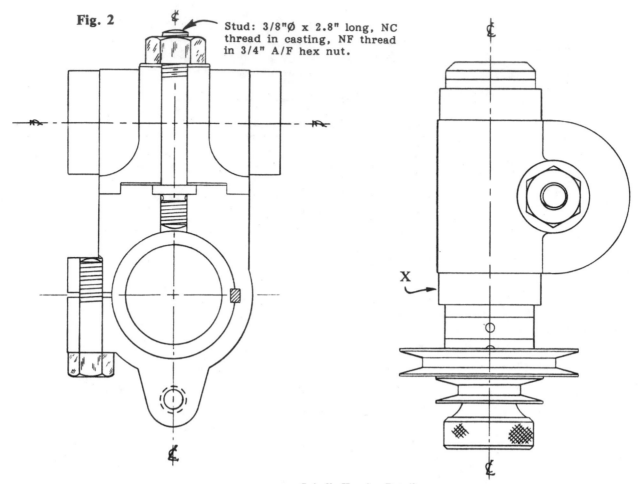

Fig. 2

Stud: 3/8"∅ x 2.8" long, NC thread in casting, NF thread in 3/4" A/F hex nut.

X

Spindle Housing Details

The internal details of the Milling Spindle can be pretty much the same as in the Milling & Drilling Spindle detailed earlier in this book, although more sophisticated bearings would be appropriate if the unit is to be used as a grinding spindle; as noted in the text, preloaded angular contact ball bearings would be best.

The means of mounting grinding wheels to the spindle should be given some careful consideration. A practical arrangement would be to put one's wheels on taper shanks (use a self-releasing taper - which see elsewhere herein) held in with a draw bolt.

See next page for Jockey Pulley and Arm details. The Jockey Pulley Arm would be clamped to the turned area marked "X" in the right hand drawing above.

— — —

directions. Any spot missed going in one direction would have been fairly sure to have been caught in one or more of the other three directions. Any little high spots remaining would be knocked off pretty quickly in the lapping operation, which is next.

Let's pause here for a minute to consider what we've done, and what we've got from it.

As a starting maneuver, we've lapped one face flat, though we haven't said much about how just yet. We've found, by means of a sensitive dial indicator, that the Block is not perfectly parallel. We've put the Block on a very flat plate with the out-of-parallel face uppermost. We've slid the Block under a grinding wheel in each of 4 orientations, so that the out-of-parallel face has had 4 sets of cuts taken over it. Any previous lack of parallelism between the top and bottom faces should by now have been pretty much eliminated.

Now, at this stage, depending on what sort of co-ordination you possess, or how fussy you care to be, the surface will have a less or more uniformly spotted appearance: - it has been "spot ground", and the process described up to this point is called "spot grinding". And if you think that the appearance of a spot ground surface might be attractive, believe me, it is eminently so.

But its flatness can be further improved by lapping.

Fig. 3 Jockey Pulley and Arm Details

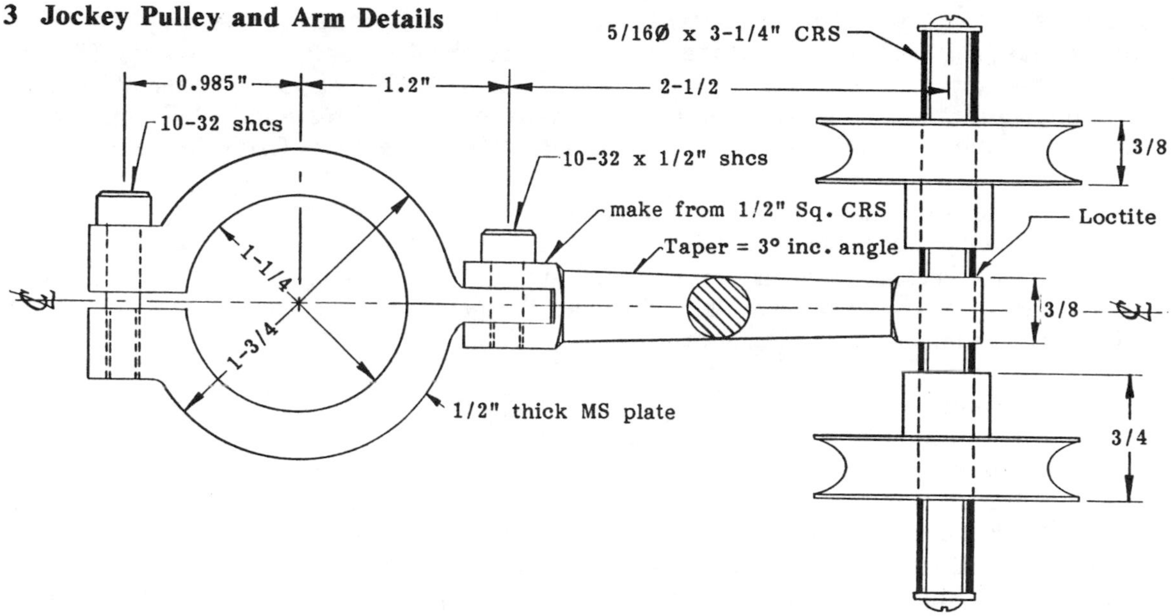

5/16Ø x 3-1/4" CRS

0.985"   1.2"   2-1/2

10-32 shcs

10-32 x 1/2" shcs

make from 1/2" Sq. CRS

Taper = 3° inc. angle

Loctite

1-1/4

1-3/4

3/8

3/8

3/4

1/2" thick MS plate

The Lapping Equipment

Jake had 3 cast brass lapping plates. See Fig. 4 for details. He squirted a little water onto the main lap, and then went over it with the ring lap and then the plug lap. Lapping compound from a previous use immediately showed its presence, and Jake said the main lap seemed to be low in the middle today - evidence of this was the brighter appearance of the lap's face in a band about 2" wide around the periphery, and a darker appearance in the center. He applied the plug lap, and made a pass or two around the main lap, moving the plug in smaller circles as he did so. Satisfied, he took all three laps to the laundry sink and scrubbed them off with a scrubbing brush.

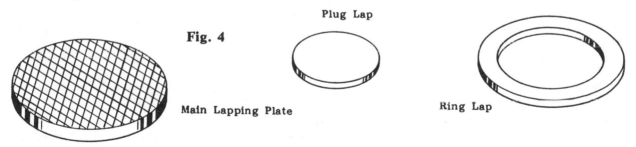

Plug Lap

Fig. 4

Main Lapping Plate

Ring Lap

Back in the lapping department, he spooned a small quantity of very fine aluminum oxide powder - maybe 1/32nd of a teaspoon or less - onto the main lap, sprinkling it about over the surface. Another few squirts of water, a pass or two with the ring lap, and the plug lap, and he pronounced it good.

We put the spot ground face of my Block down on the main lap, and proceeded to lap it flat, moving it in overlapping circles, back and forth, side to side, and so on. The lap cut surprisingly quickly, and in a very little time the spot ground face of my Block, when turned back up to the light of day and wiped clean, showed a largely lapped appearance, with much - but not all - evidence of the spot grinding gone. We lapped some more and then took it out for another test with the dial indicator. This time the two faces appeared to be parallel, within the sensitivity limits of the indicator.

Jake brought out a 1"x3"x12" steel parallel, one of a pair, hardened, ground and lapped true. My Block was set on this parallel with its newly-lapped face down, and we checked the two long sides for squareness with a gage maker's square.

El Squaro Supremeo

Let's digress for a few moments and have a look at Jake's square. See Fig. 5. This little laboratory-grade square had been made as one of a batch of several, and given to Jake early in his days as a gage maker during WWII in England. It had been lapped all over after hardening and grinding, and was as square as it could possibly be made. We handled its mirror finished surfaces with a piece of chamois leather, to insulate it from the heat of our fingers.

Such a square is not something one could readily make for oneself. It would have to be made as one of several, all lapped to flatness and size as a group, and the end of the blade abutting the beam lapped square to the sides in a fixture made for the purpose.

Fig. 5

If you didn't want to make a Master Reference Square, as described earlier herein, you could buy the most accurate bevel-edged hardened and lapped square available - laboratory- or inspection-grade if possible. Starrett, Brown & Sharpe, Mitutoyo and others make such squares, but good ones are not cheap. Or you could buy a shop grade square and work it over yourself; there is enough info in the section on making a Master Square that no more than that need be said here.

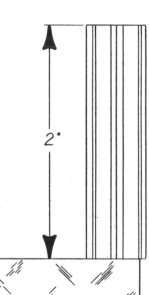

TIP: When using a square, do not force it into hard contact with the work, in hopes of making sure you get a good reading - you won't, and you'll be none the wiser. If the angle being measured is more (or less) than 90°, such that the far end of the blade touches the job, and you can see daylight between the job and the inner (or outer) corner of the square, if you press the square hard against the work you will spring the square out of true, and mess up. It is possible to spring the tip of the 2.5" x 0.094" blade of a top quality 24" hardened square out of true by 0.001" for each pound of force applied.

Therefore apply the square but lightly to the work; if it gives a "black out fit", you will know it. If it does not, you will know that too, and you will be wiser.

The old gagemaker's super-accurate square

Well, we had a super accurate square to work to, and we had a Toolmaker's Block that had had a $40 surface grinding job in a tool and die shop. We'd lapped the two biggest faces parallel. Would the sides be square to the two big faces?

They were not! We could see light between the square and the Block on both side faces. We decided to square up the face that was closest to being square already, so back we went to the lapping department.

Jake attached a small magnetic chuck - just a little fella about 2-1/2 x 2-1/2 x 4" - to the big face of the Block on the side he wanted to bring square (see Fig. 6) and we lapped a little. The weight of the magnetic chuck caused the lapping effect to be greater on that side of the Block.

After lapping briefly, we cleaned the Block and checked it for square again. Not quite square yet, so back to the lapping plate. In a couple or 3 tries, we got it square. So we now had 2 big faces parallel, and one side between square to them both.

Note that we did not spot grind the 3rd face. What we wanted to do was to bring it square to the first 2 faces, not make it parallel with the opposite face, which we knew was also not square to the 2 good faces. If you don't see this immediately, think about it as you read the next paragraph.

The next step was to spot grind the face opposite the newly squared face. This we did. Then we lapped this fourth face, and checked it for square. It wasn't quite perfect, so Jake said lap it some more, but put more weight on the one side of the Block - by this time he had me doing the lapping. It didn't take much more to get that side so square with the two big faces that it was a blackout fit against Jake's "Squaro Supremeo".

How about the ends? Turned out they were pretty near perfect, so we gave them each a bit of a turn on the lap, just to smooth them up, & checked 'em again - a blackout fit! Time to clean up.

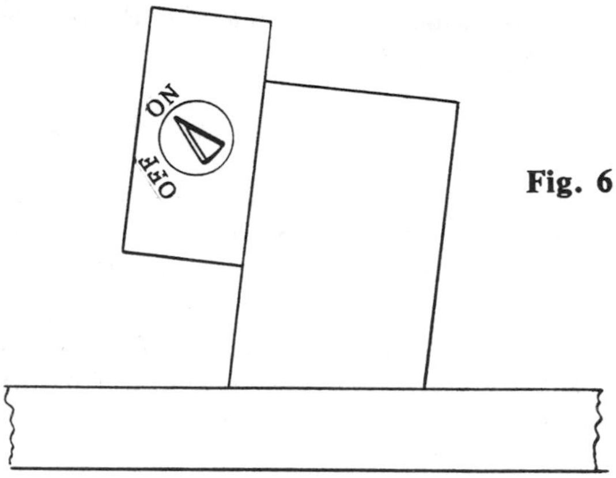

I gave the Block a good scrub with soap and water that evening, mopped the water out of 70 or so little holes, (Faster: go over it with the air hose), gave it a shot of Starrett M1 oil, and there was my Toolmaker's Block, lapped square and flat within a few hundred thousandths of an inch, according to Jake.

Fig. 6

Weighting one side of a workpiece with a small magnetic chuck to help lap out an out-of-square condition.

—— · —— *//* —— · ——

Lapping is Probably Superfluous

Since writing the above, I have done some further reading in various sources, and would add the following thoughts:

1. As noted above, the lapping operation is probably superfluous, for most work - i.e. spot grinding alone, if carried to the point of "sparking out" over the entire surface, is probably enough for most purposes. Lapping is only needed when making gages for measuring the very finest of tolerances, or where an exceptionally high finish is wanted, e.g. when the surface should be good enough to wring to a gage block.

2. Advice on the type of grinding wheels to use can best be had from the maker's representative - they are the experts, and they will have good literature available. (If they don't, buy from someone else!) It is likely that a 54 to 60 grit wheel will be found suitable. Wheels of finer grit will not necessarily give finer finishes, and in fact may glaze or clog more readily, hindering your efforts. The wheel must cut freely, and must exert only the smallest possible force upon the work. To achieve this, the wheel must be of the right grade, be properly dressed, and be used to take light cuts only. Any force big enough to cause distortion in the workpiece or deflection in the grinder will lead to inaccuracies.

3. The wheel must be kept in perfect cutting condition through proper and frequent dressing. For rough grinding, the diamond dresser should be traversed across the wheel face rapidly; for fine finishing work, a slightly slower passage of the diamond will leave the wheel somewhat smoother - it will cut less quickly, and leave a finer finish.

4. As a general rule, when dressing a grinding wheel, do not dawdle: get the diamond across the width of the wheel in less than a second. If you dawdle, you will leave the wheel face burnished rather than sharp, completely defeating the purpose of the exercise.

5. Dressing the wheel will be a frequent requirement, and for this a diamond dresser on a shank, mounted sticking upwards in a block of steel, can be used in conjunction with a steel straight edge used to guide it in a straight line pass under the lowest point of the wheel.

6. Dress the wheel at the first sign of glazing. Longer intervals between dressings is false economy, as the wheel cannot then produce good work, and its productive life will in fact be shorter. The diamond dresser should be of the size recommended for the size of stone to be

dressed - this info too will be available from the abrasives people - and it should be a good quality diamond. A diamond should be re-set before it becomes too dull; if this is done, it will last a very long time.

On the matter of re-setting a diamond

I don't know where one could get this done, but the logical first place to try is the outfit that made the diamond dresser in the first place. Contact them and see if they are interested in doing it for you. Or....

See also "Setting diamonds for wheel dressers," in the index to your copy of the *American Machinists' Handbook*: I have a 7th Edition (1940) copy of same - at page 494 therein are drawings of 3 ways to set a diamond. Any of the methods shown could be carried out by any careful hsm.

(I was most interested to see an article by Jim Hesse on diamond lapping etc. in the Jan/Feb '88 issue of *HSM*. I wrote to him to ask if he'd care to expand on what he said about resetting diamonds in diamond dressing tools in the concluding paragraph of his article. What follows is based on what he had to say in the original article, and his reply to my letter:

"The diamond will be found to be secured in the end of the dressing tool's shank by the metal of the tool shank having been crimped over, or by spelter (brazing alloy) in an oblong hole.

"Heat the spelter to permit removal of the diamond which is to be re-set. Drill a new hole slightly more than big enough to take the stone, and about 3/4 the depth of the stone. Make this hole oblong (out of round) by squeezing the drilled area in a vise. Flux the hole, fill with molten spelter, and re-insert the stone, pushing it into the molten spelter with a carbon rod having a small dimple in the end, or put the stone in place with a pair of tweezers. Re-machine the end of the dressing tool shank in the lathe, and you are in business again."

Mr. Hesse pointed out, in answer to one of my questions, that diamonds can take considerable thermal shock. Proof of this is that the tip of a diamond will be incandescent at the point of contact when dressing a grinding wheel, while most of the rest of the stone will be about room temperature, as is the dressing tool shank.

7. If the grinding spindle is good, and the mountings for it are solid in comparison with the cuts being taken, the work will come out with a flatness tolerance as good as, and possibly better than, the surface plate on which the spot grinding is done. An accuracy of ±0.000,02" (20 millionths of an inch) is possible if the last cuts are sparked out completely.

8. Spot grinding should be done with cuts about 2 or 3 tenths deep, this being reduced to 1 tenth for finishing passes. Half a thou should regarded as a heavy cut, and not exceeded.

9. If you have access to a surface grinder, you can put a good surface plate on same, locking out the normal table movements, and go to work - obviously the spot grinding *technique* is the thing, not the surface-plate-plus-pillar-in-the-corner arrangement. Apparently, after the first two passes under the wheel have been made *against* the push of the wheel, subsequent passes *at that wheel height setting* can be made in just about any direction.

10. If the workpiece weighs less than three pounds, or is too small to be securely hand held, it is liable to be lifted off the surface plate by the drag of the grinding wheel. In such cases it is best to stick it to a small magnetic chuck, the latter also having been previously spot ground top and bottom.

11.   The rise & fall arrangement for the grinding head must have a nice action, so that its movements are reliable and predictable: if it is lowered 2 tenths, it should come down by about that amount and not five or ten times it. This means that you will need to make or buy a well-fitted and smooth working slide, and a ditto screwfeed.

12.   Eventually the surface plate will wear down somewhat in the most used areas.   It can be returned to its maker to be lapped flat at a nominal cost. Granite plates will have a much longer life than would a cast iron plate used for this sort of work; I am guessing, but I suspect that a hsm would be unlikely to wear it out in a lifetime. If you really wanted to get fancy, you could have a grid of small grooves cut into the working face of the plate, say 1/16" grooves on 1/4" centers. This gives the grit somewhere to go, and helps reduce friction between the job and the plate - easy movement of the job on the plate is essential to success with the spot grinding technique.

13.   Grinding can be done wet or dry. Dry grinding is more common.

Made Flat by Scraping

One could come at the whole matter of producing surfaces that are accurate for flat, square and parallel from an entirely different direction.   You have undoubtedly seen scraped and frosted surfaces on machine tool ways and elsewhere. Scraping is another way to produce highly accurate surfaces. There are two books you should see, if this interests you. Both are available through your local library, either off the shelf or through Inter-Library Loan.

1. *Foundations of Mechanical Accuracy*, by Wayne R. Moore; published by The Moore Special Tool Company, Bridgeport, CT. This book details how the Moore people build extreme accuracy into their jig borers, how they scrape the machine ways to master gages, design of master gages for greatest stability, etc. It is a most interesting book from which to gain an overview of the whole matter of high accuracy work. Because of its cost – about $70 at time of writing – you will likely content yourself with looking at a library copy.

2. *Machine Tool Reconditioning, and Applications of Hand Scraping*, by Edward F. Connelly; published by Machine Tool Publications, Suite 208, 1600 University Avenue W., St. Paul, MN 55104. The first 1/3 or so of this book might well be titled *The Art of Hand Scraping*. If you want to know how to produce flat, true surfaces by hand scraping, this is the book to have. It is about $98, but in spite of that, you might want to own it, once you've seen it. There is probably little to know about hand scraping that is not covered in this book.

[Added to the 7ᵗʰ Printing, January 1999: For an introductory demonstration of hand scraping methods, you might want to have our newest video, **Examining a Used Lathe and Milling Machine – A Machine Tool Rebuilder Shows How**. See page 206 for a complete description of this video.]

I would say that a man with patience and the desire to do so, could, with the info given herein, and in the above two books (and the above video), produce for himself straightedges, squares, angle plates, true squares, and similar high precision tooling and accessories* in which he could have the highest confidence, to say nothing of much pride and satisfaction at having made them himself.

* Some of these items could with advantage be bought as ready-made, standard workshop-grade commercial products, and then worked over brought to very high standards by application of the techniques outlined herein. (Of course, to merit such refinement, they must be well and soundly made in the first place.)

Final Note: At pages 117-122 of this book, you will find an article on sharpening straight razors and other fine edged tools. In it you will learn how to make an exceedingly fine abrasive from charred rye straw. I don't know how it would work for lapping purposes, nor how fine it would be, but I draw your attention to it here, just in case you might otherwise fail to make the connection.

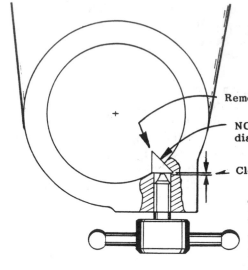

Remove sharp edges of Key

NOTE: This face should be truly on the diameter of the shaft - i.e. truly radial.

Clearance here is quite permissible

The triangular gib key is an element of machine design which you will not likely find in most other sources. It is useful where a part slides along a round column, and must be kept in radial alignment when locked, and must be free of play when sliding. The working arm of a radial drill would be an example of where it might be used, although other means of achieving the desired end are used in that instance.

The triangular gib key was invented in the 1950's by one David Urwick. He patented the idea, and described it more than once in *M.E.*; although I cannot locate his first mention of it, the last mention was in the August 15, 1980 issue, at page 1013; at that time he said the patent had long since expired.

A logical use for it arises in the construction of something like the Potts Milling Attachment shown a few pages earlier, although there is no pressing reason to include it, or to want to ensure radial alignment, if you are making the Potts unit just for surface grinding purposes.

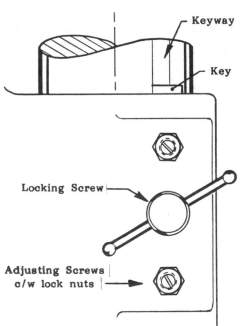

Keyway

Key

Locking Screw

Adjusting Screws c/w lock nuts

In the early 1950's Mr. Urwick built a small universal machine tool which he dubbed "the Metalmaster" - an apt enough name, too, as it was a small variable center height lathe, horizontal boring mill and milling machine, all built into one very neat package. He incorporated his triangular key in two places in the Metalmaster, and used the finished machine up until his death in 1986.

Incidentally, castings and drawings for making Urwick's Metalmaster are available. If interested, **phone** me and I will tell you how to go about getting a set. The Metalmaster had a number of interesting features, not least of these being that the center height could be varied, as noted above. The main bed was arranged much like the knee of a milling machine, and was kept in radial alignment on the round column of the machine by Urwick's triangular gib key, while the tailstock travelled on a separate bed not unlike the overarm of a horizontal milling machine. As a model engineer's machine tool, or as an adjunct to a regular lathe, Urwick's Metalmaster has a lot to offer. Lotta work to make one, of course.

The drawings at left above show the nature of the gib key and from them you will be able to work this design element into your own projects.

The big question is: "How to cut the keyway in the bore?"

There are only a few ways:
1. with a shaper or a planer or a vertical slotter;
2. chip it out with a cold chisel and finish it by filing; few men today have the skill to do this.
3. broach it with a special broach.

Of the above, the first is by far the best choice, so long as you have access to one of these three machine tools. A single point tool could be set up in a good-sized vertical mill, and the quill worked up and down by hand to shape out the cut - not fun, or fast, but would probably get the job done.

The use of the triangular key by readers of this book would, I think, please Mr. Urwick, for he was - and rightly - proud of his invention.

Pretty Good Shop-Made GAGE BLOCKS

The least expensive set of gage blocks will set you back $150/200 for a 36 piece set. For almost any hsm's activity, the toolroom grade would be more than adequate, and in fact few if any of us will ever have any need for a set of gage blocks. There are a couple of less expensive alternatives, both of which are practical (i.e. one would have valid uses for them) and within the reach of most any hsm.

You can make or buy a set of "shop blocks", also called "space blocks". These, as commercially made, consist of 36 pieces, 3/4"Ø, drilled and tapped 1/4-28 at center, hardened, and lapped to thickness to within a tenth of a thou. My set came from Travers Tool.

All things considered, a bought set, at $60/70, is probably a tough deal to beat, and will take less of your time to earn than make. But if a fella put his mind to it, he could make a pretty fair set of these for himself, aside from the hardening. If they don't need to be accurate to ±0.000,1" for your purposes, the job becomes that much easier. Here's how I would go about making a set:

Face off the end of a bar, and part off a slab a little thicker than wanted for any given piece in the set. You might then want to drill & ream 1/4"Ø, in place of the tapped hole in commercial ones. What you do next will depend on the degree of perfection you desire:

1. You could just clean up the faced ends with a hand file, and mike the daylights outta them.

2. You could surface grind, or spot grind and lap, to spec. (If the latter, stick each piece to a carrier block of suitable dimensions, previously spot ground and lapped on 2 opposite faces. (How to hold a disk 0.050" thick without a magnetic chuck? Use "Superglue" or double-sided carpet tape.

   (Full info on both these techniques are given in the plans for Lautard's OCTOPUS, available with or without a casting, direct from your favorite author. Write for details. GBL)

Finally, check for size under a good sensitive dial indicator (see below), and/or with a 0.000,1"-reading mike. (The latter is not a bad thing to have, but the former adds greater versatility to your equipment, in that you probably already have other mikes, and if so, a tenths D.I. will provide more "new fun" than a tenths mike.)

Obviously, if making your own space blocks, you can make them any size you want or need. You can calibrate (measure) them, and not have necessarily to lap them to perfect nominal size.

If you know the sizes of your blocks, and among them have a suitable combination of blocks to make a desired height, there's no need that they be perfect for size, although "off" and odd sizes may result in a needless proliferation of pieces.

Etch the size on the finished block, or stamp it on at an early stage, dress down the stamp impression with a hand file, and then lap to size, leaving only a minimal stamp impression showing on the finished piece. In the set I have, blocks under 0.090" thick have the size etched on the working face, while those 0.090" or thicker are stamped on the side.

Note that the two faces of each block must be parallel - i.e. they must not form a wedge. This can be checked with the tenth's indicator - see the section on Spot Grinding & Lapping for more info on this.

MAKING AND USING TOOLMAKERS' BUTTONS
by Mac Mackintosh

Toolmakers' Buttons are an accessory not often found in the amateur's shop. This is unfortunate, as they are a useful aid to precise location of holes - e.g. the location of holes in the side-plates for a small gearbox. A first class set of Toolmakers' Buttons can be made in about an hour, so if you want a set, the lack thereof is easily corrected, and they will be found useful forever after when doing precision layout work.

Toolmakers' Buttons are simply thick-walled cylinders which can be secured to a flat surface by means of undersized screws. They can be bought from most tool suppliers and are usually sold in sets of four, one a little longer than the other three. They come in various sizes. Small Buttons may be 1/2"Ø, with 3 at 1/2" tall, the 4th one 3/4" tall. A larger set of Buttons will be 3/4" or 1" diameter; for the 1" size, the lengths may be 1" and 1-1/4".

(In my neck of the woods a set of 1/2" Toolmakers' Buttons from a big-name maker of machinists' tools will set you back about US$30, at date of writing (mid 1988). Thus, since making a set is not a big job, there is good reason to consider making them for yourself. GBL)

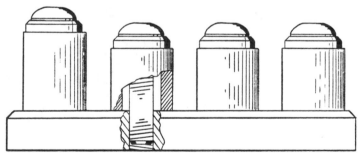

A Set of Shop-made Toolmakers' Buttons
An hour's work; a lifetime of utility thereafter

How to make them

To make a set of 1/2" Toolmakers' Buttons, start with a piece of 1/2"Ø drill rod and mike it to ensure that it is "on the nose" for size, and test it with V-blocks and indicator to see that it is not "triangular".

(Drill rod is made by centerless grinding. In the early days of centerless grinding, there was a tendency for the work to be ground "out of round" - nowadays this is very rare.)

Rough cut three pieces 1/2" long and one piece 3/4" long. Put each piece in turn in the 3-jaw chuck. Face one end, reverse, and face to length, ±10 thou. Next, drill each one 5/16"Ø full length, and break the sharp edges.

Next, from 1/2" CRS, make four 15/32"Ø washers, about 3/32" thick, nicely beveled on one side, and with a clearance hole for a 10-32 screw. Buy three 10-32 Button Head cap screws

7/8" long and one ditto at 1-1/8" long, and the set of Toolmakers' Buttons is completed.

To prevent loss, drill and tap four 10-32 holes down the centerline of a piece of 5/8 x 1/4 CRS about 2-1/2" long, and to it screw the four Buttons and their washers for safe keeping.

How to use them

Say we want to make a small gearbox, and we have made the box itself from say 3/16" steel plate. For nice meshing of the gears, we would need to lay out the locations for the gearshaft centers to a tolerance on the order of ±0.002", but even with a #1 center drill and magnifier, we are unlikely to do better than ±0.005". We could make the layout by means of a height gage and surface plate, and it will be better, but still not perfect in the final analysis, due to errors in "picking up" the intersections of the layout lines.

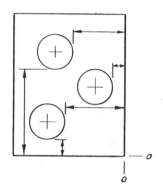

On a job such as this, Toolmakers' Buttons enable precise location of hole centers for machining.

We can instead make a rough location of the ends of the gear-shafts, using the height gage, and then drill and tap 10-32 at these points. We then lightly screw a Toolmaker's Button to the job at each location. Because the screws are undersize to the holes in the Buttons, the Buttons can be moved sideways by small amounts about their screws, permitting the Buttons to be perfectly located, with the aid of gage blocks, parallels, micrometers, angle-plates, etc. Once a Button is properly located, its screw is fully tightened.

For a final check, we would measure the distance between adjacent Buttons, either with an outside or inside micrometer, being careful to add or subtract 1/2" to allow for the diameters of the Buttons, and thus check the center to center distances.

By such means we can hold the ±0.002" tolerance required, and, with care, cut it in half.

(NOTE: If a job should happen to require the locating of more than four centers, we can do the job in stages; the use of Buttons gives positive location, and once a Button is located and screwed up solid in place, it is independent of other locations.)

The workpiece, with the Buttons in place, is then transferred to the vertical milling machine and each Button in turn is centered accurately by means of a dial indicator mounted in the spindle. (Or see "Osborne's Maneuver", at the end of this book. GBL) Once located, the Button is removed and the hole started with an end mill (remember that there is a tapped hole here which would upset the location of a center drill or pilot drill), and then brought up to size. In some cases the holes would instead be tooled out with the job set up on the lathe faceplate.

In operating a milling machine (or even a jig-borer), one of the gremlins to be guarded against is cumulative error. Even with a new mill of good quality, you are not likely to be able to hold ±0.002" tolerances over a series of holes if you rely on the co-ordinate motions of the machine. Using Toolmakers' Buttons, you can locate your holes and check them against one another before you begin to drill the holes. As noted earlier, it is not too difficult to hold tolerances to ±0.002", and if you have a good "feel" with a micrometer, it is even possible to work to ±0.001" (though here you will have to exercise a lot of care, however good your "feel").

Making a gearbox is only one of the applications that Toolmakers' Buttons can be put to. There are many other uses for them and even if you don't have any immediate work in which they could be useful, it is worthwhile to make up a set to have on hand when needed.

THE OLD CIGARETTE PAPER TRICK

A classic piece of trickery employed by tool & die makers is the cigarette paper test. This is probably as good a place as any to insert a couple of comments about it, as it may someday be useful to you.

Suppose we have a parallel bolted to a faceplate, or to an angle plate, or similar, and we want to set up a second parallel exactly at right angles to it. We might want such an arrangement as a pair of "locating fences" for a jig plate, or whatever.

We get out our Reference Square, and set it on a little silk pillow. When the second parallel is bolted in place just finger tight, we use the Square to set it square to the first. When we think they are properly set, we tighten up the bolts somewhat, and insert two cigarette papers between the second parallel and the blade of the Square - one near the corner, and the other near the end - and press the Square into place hard enough to just nip the papers. We then try to pull out first one paper, and then the other. The "feel" of the pull required to move either paper should be the same.

We can have a double check of this setup by reversing the square and repeating the test.

A second trick of this same type can be used to test the flatness of a workpiece against a surface plate. We would put the job on a surface plate with 4 cigarette papers under the corners of the job (assuming it has 4 corners), and we would then try to pull out each paper in turn. The force required to budge the paper at each corner will be the same only if the 4 corners lie in a single plane. Now the job may be flat, but it may not be, either - it could have a dished center; but you will understand the idea, and somewhere down the road you may find a use for this idea, or some offshoot of it.

A COMMON SHOP MATH PROBLEM

(Many readers will know how to handle this one already, but there may be some who don't. It will also serve as an introduction to some other ideas you may find interesting, so bear with me.)

Consider the triangle at right.
Problem: Find Θ if AC & BC are known. Or, to put it in more down-to-earth terms, "If you want to put a bevel on a job, either on paper, or in metal, what's the angle involved, given that the bevel goes from "A" to "B"?

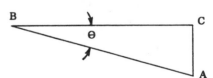

Consider the tip of a toolmaker's clamp:

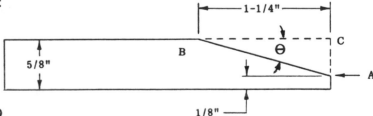

Our problem is to calculate the angle Θ

The first thing we see is that AC = 1/2"

Then Θ is the angle whose tangent is AC/BC, i.e. 0.5/1.25 = 0.4000.

"So, the angle is 0.4°?" you say.

No, Θ is the angle whose tangent is 0.4000,

therefore Θ = 21.80°

which is to say that if we look up the tangent of 21.8°, we will find that it is 0.4000.

Now all of the above can be expressed more succinctly, thus: "$\Theta = \tan^{-1} 0.400$"

which is said: "Θ equals tan-to-the-minus-one of 0.4".

Some people say "arc tan" instead of "tan to the minus one", but both mean the same thing.

Now maybe, like me, you had a little exposure to a book of trig tables in high school, and you hated it. Well, take heart, things ain't what they was when we was in school. Buy yourself a pocket calculator with trig functions built in, and you'll have the subject by the tail on a downhill pull. Suitable calculators can be had for $10 and up. My own is a Hewlett Packard HP11C - I've seen these on sale for $40 Stateside, in early 1987.

Trig is an indispensable tool for machinists. It is an interesting and useful branch of mathematics, whose fundamentals are easy* to master, and whose more esoteric elements need not be learned, because they are always at hand in any trig handbook. If you lack, and want, a knowledge of trig, you will find it well covered in (1) *Machinery's Handbook*, (2) *American Machinist's Handbook*, and (3) *Wolfe & Phelps' Mechanics Vest Pocket Reference Book*.

*For sure some reader is going to think "Easy? I took trig in high school, and I never understood it (or passed it)!" Well, I had the same problem - but now there's a motive for learning it (you can see where you can use the knowledge) and you are older and smarter now, and finally, those little pocket calculators have all the info stored in 'em, so that looking up sine, cosine and tangent values in a table is a thing of the past: you key in the angle, push a button, and there's the sine of that angle, before you can blink. Want to see what the cosine of the same angle might be? Push a couple other buttons, and you've got it. Easy? Yup.

From here on, the assumption is made that everybody who reads this book has a working knowledge of trig, which is probably true.

LATE FLASH!!

Just as I was putting the finishing touches on the last few drawings for this book, I spotted an item called a SIMPLE FYER. This you gotta have! It is good, cheap and simple. It is a pocket size (3-1/2" square) dial type gadget giving all the trig info most of us will ever need to know. You just turn the dial until the part of the triangle you want to solve for shows up in one little window of the SIMPLE-FYER, and the appropriate formula to find it is sitting there looking back at you in another window. What more could a lazy old machinist want? About $4, from Reid Tool Supply Co., whose address will be found in the Appendix.

TAPERED SIDES & RADIUSED ENDS:
ANOTHER COMMON SHOP MATH PROBLEM

Building on the foregoing angle problem, here's another one, whose solution may not be quite so obvious, which comes up not infrequently: A discussion of it will lead us to another topic, also of interest.

What is the angle from centerline made by the line joining two arcs of different radius?

Consider the drawing at right: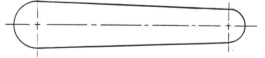

This might be encountered in making a crank arm, like so: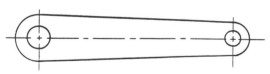

We might be milling this out of a piece of say 1/4 x 3/4" CRS, so we want to cut the two radii and mill the taper between so everything blends in nicely; we'll probably insure the latter with the aid of a few finishing strokes with a file.

My approach would be to first lay out the centerline and the hole spacing, and scribe both arcs and the lines showing the tangency points. Then I would tool out both holes. These might be the same size or two different sizes, one or both might be reamed, counterbored, bevelled, countersunk, tapped, screwcut, or whatever - these extras would dictate whether I'd do it in the mill or on the lathe with faceplate or 4-jaw.

Next step would be to mill the larger radius of the two. This would be done on my 3-3/4"Ø rotary table, or by multi-facet milling* followed by filing. Then set up with one of the straight sides (parallel to centerline) tilted at appropriate angle in mill, and machine off excess. I'd likely set up the angle via a vernier protractor if I wanted to be fussy, with an ordinary protractor if not fussy at all (unlikely) and if I wanted to be as close as my best tooling and efforts allow, I'd use a sine bar. The job would go back on the rotary table to machine the smaller radius, after which a few strokes with an 8" hand file would finish the job, and I can assure you, it would look real fine.

*a technique described later herein

Three things come out of this that are of interest.

1) What is angle Θ, (i.e. how do we work it out?)

2) What is the arc of each radius? (i.e. what is Θ_2 and Θ_3?)

3) How do we set up a job at a desired angle to very close limits?

Problem #1: Calculating Θ
This is required by both the draftsman and the machinist - in many cases these two may be one and the same person. If not, and if the draftsman omits this info from the drawing, the machinist will not be very happy. How to do?

Consider Fig. 1

To machine this part on a rotary table, we will want to know Θ_1, Θ_2, and Θ_3. Obviously, R_1, R_2 & L will be known.

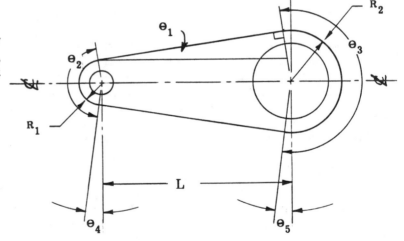

Fig. 1

Fig. 2

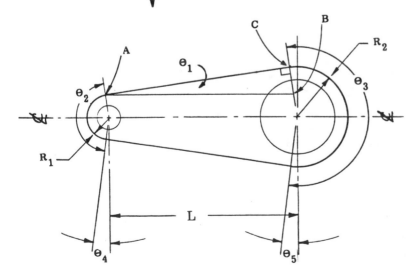

We'd better throw some letters at Fig. 1 so we can discuss it like adults. Therefore, see Fig. 2, at left, whereon let us also agree that for our example,

$$L = 4"$$
$$R_1 = .75", \text{ and}$$
$$R_2 = 1.25"$$

74

Consider the triangle ABC

We know:

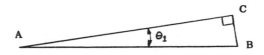

$$AB = L = 4"$$

$$BC = R_2 - R_1 = 1.25 - 0.75" = 0.5"$$

and angle ACB is a right angle.

Then, $\Theta_1 = \sin^{-1} (BC/AB) = \sin^{-1} (0.5/4) = \sin^{-1} 0.125$

$$\Theta_1 = 7.180755° = 7° \ 10' \ 50"$$

That info will allow us to mill the taper into the edges of the piece of CRS we are likely going to make this part from. The next stage will then be to machine the radius at each end, which brings us to....

Problem #2: Calculating the arc of each radius

A little thought will show that $\Theta_1 = \Theta_4 = \Theta_5$

(Similarly, $\Theta_3 = 180° + 2(\Theta_5)$

$= 180° + 14.36° = 194.36°$, say 194°)

Therefore $\Theta_2 = 180° - 2(\Theta_4) = 180 - 2(7.18075)$

$$= 180° - 14.3615° = 165.6385°,$$

or...in bald round numbers, say 166°

(We certainly can't lay out or file the arc to closer limits than that. If using my small rotary table (see p.193, **TBMR**), I'd set the stops to give me a table rotation of 166°, so why call out the angle to 3 decimal places? Excessive precision is foolishness - if you can't produce it, why put it on the drawing?)

Problem #3: Setting up a desired angle to close limits

Say we did in fact have a need to produce a desired angle to very close limits - what to do? This brings us to the matter of...

THE SINE BAR

Say we need to produce a desired angle to very close limits - what to do? Use a sine bar. Such a device is probably better bought than made, given the kind of precision we may be after if we need a sine bar, but let's look at the sine bar principle, and in case you're thinking about making one, I'll point out something to help keep your tail outta the cracks if you do make one.

I suspect few will need any introduction to the concept of the sine bar, but for those who might, here it is:

Say you have a workpiece that requires to be machined off at an angle of precisely 12° 43' 26". Say job is 1/8" thick, and starts as 1" wide.

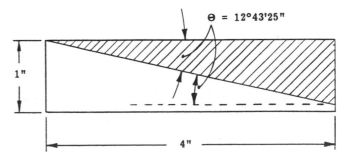

How to do this in a civilized and easy manner? Clamp the job to an angle plate at the correct angle - as shown on the drawing at the top of the next page - and mill (and possibly grind, and/or lap) the unwanted material away. Job done. The question, of course, is how to set it at precisely the angle desired. The obvious answer is to raise one end until the part is tilted at exactly 12° 43' 26". In our trade, this business of raising one end of the job the right amount is called the sine principle. How to do it? Easy, but first let's see how this works...

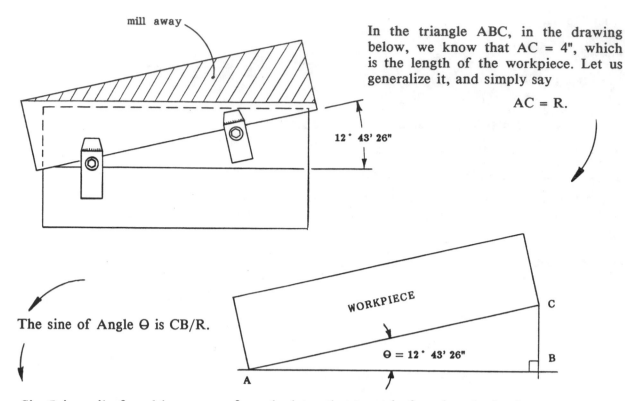

In the triangle ABC, in the drawing below, we know that AC = 4", which is the length of the workpiece. Let us generalize it, and simply say

$$AC = R.$$

The sine of Angle Θ is CB/R.

Sin Θ is easily found by means of a calculator that has trig functions in it, thus:

12° 43' 26" = 12.723,888,89° (we convert it to decimal degrees first)

then:

sin 12.723,888,89° = 0.220,252,924

Now, we know sin Θ = CB/R, and we know both sin Θ and R, so we've got 'er licked, thus:

CB = ? = R(sin Θ) = 4" x 0.220,252,924 = 0.881,011,698"   ... or say 0.881,012"

Now, you will see a problem: how do we go about raising one end of the job by this amount, and if we try to do it by putting a pile of packing (e.g. gage blocks) under the end, how to get them right at the end, i.e. thus ⌐

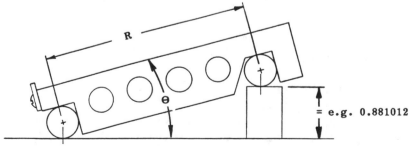

and not thus ⌐

The SINE BAR provides a simple solution to this problem.

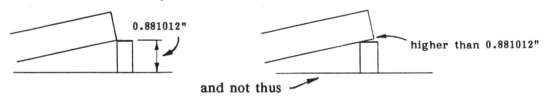

Here, R = the distance between the rolls of the sine bar, usually some convenient whole number like 3.0", 5.0", or 10.0". R must be known accurately, i.e. it will be precisely 5.0000" ±0.000,1" or so.

TIP: If you are making a sine bar, make it as at A below, not as at B. If in making the distance R to some precise dimension, you take too much off at X, so that R is decreased by say 0.0016", you can correct this by going back to the other end and taking some more off at Z. If made as at B, and you get into the same jam, there's no way out!

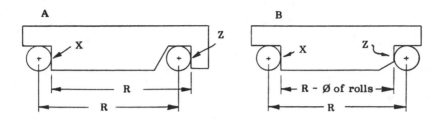

Now, how do we use all this stuff?

One way is to set the job up on the surface plate:
1) Raise one end of sine bar on space blocks to desired angle.
2) Put an angle plate behind sine bar.
3) Put job on top of sine bar, (job thus sits at the angle the sine bar is set to).
4) Clamp job to angle plate.
5) Take job and angle plate (but not sine bar) to mill, and...
6) Machine to spec.

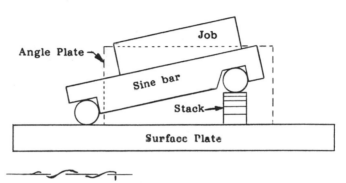

A SIMPLE FLAT SINE BAR

You can make an accurate flat sine bar from a 1/4 x 3/4 x 12" piece of ground flat stock. (Maybe we should call it a sine blade.) Drill and tap 10-32 at various locations. Attach a pair of rollers, say 3/8" thick, made from 5/8"Ø drill rod. (Just in case you are wondering, these discs are not free to turn. Also they should both be exactly the same size.) Adjust the discs to a desired spacing, or

A cheap but accurate Sine Bar

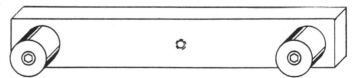

Determine center-to-center distance of rolls and mark this dimension on the sine bar. Thereafter, multiply sine of any desired angle by this dimension to obtain necessary stack height.

measure the spacing they happen to come out at, or if beyond the range of your calipers, take somewhere and get the spacing measured precisely, and record it.

NOTE: The rolls need not be proud of the under side of the bar: you can put a parallel under the low roller, thus:

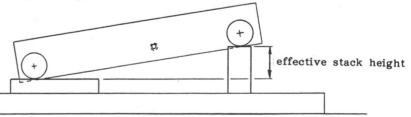

Given the means to produce an accurate stack height, the greater the distance between the rolls of a sine bar, the more accurate it will be.

77

ERRORS AND ACCURACY
in the use of a sine bar

It may be interesting to consider how the accuracy of the foregoing set-up is affected by errors or inaccuracy in the two variables which are present, i.e. R and height of the gage block stack.

Say we are using a shop-made 5" sine bar and we think the roll spacing is 5.0000", when in fact it is 5.0025". (i.e. 2 1/2 thou over.)

Say we want to set up an angle of 2° 48' 21". The height of the gage block stack needed will be

H = stack height = 5.0000 x sin 2° 48' 21"

$$= 5 \times \sin 2.805,833$$

$$= 5 \times 0.048,951$$

$$= 0.244,757"$$

But in actual fact, R = 5.002,5", as specified above, (although we may not have found this out yet). Now if our stack of gage blocks is dead on 0.244,757", we would actually produce an angle of

$$\Theta = \sin^{-1}(0.244,757/5.002,5") = \sin^{-1} 0.048,927$$

$$= 2.804,430° = 2° \ 48' \ 16"$$

The error is about 5 seconds = 5/3600, or about 1 in 720, or about 0.14%.

But now suppose we buy a sine bar upon which the gnomes at Apollo Tool have labored long and diligently, yea even unto such perfection that R = 5.000,000,000", but our gage blocks are low end Eastern Block commie-ware and our stack is short by 0.000,7".

Say we want Θ = 2° 48' 21"

H = 5.0 x sin Θ = 0.244,757", as before.

Then $\Theta = \sin^{-1}[(0.244,757" - 0.000,7")/5.000] =$

$$= \sin^{-1} (0.244,057/5.0) = \sin^{-1} 0.048,811$$

$$\text{Therefore } \Theta = 2.797,799° = 2° \ 47' \ 52"$$

The error is about 29 seconds = 29/3600, or about 1 in 124, or about 0.8%.

Now what does all this mean to the guy who buys a ±0.000,2" sine bar, and a set of ±0.000,1" shop blocks? Say he seeks an accuracy of ±1' (one minute). Can he get it?

I won't bore you with any more arithmetic. The answer is not "Yes" but "Far better." In fact more like ±1/3rd of a minute (i.e. ±20 seconds), and often better.

What's the implication? If you want/need better accuracy than a vernier protractor can give you, make or buy a 5" sine bar, (or for more "forgiving-ness", a 10" ditto) and a set of shop blocks. With these you can set up any desired angle to a level of accuracy about 10 times better than with a vernier protractor.

> NOTE: The significance of inaccuracies in a sine bar set-up increases as the angle increases. If the desired angle exceeds 45°, the sine bar becomes rapidly more sensitive to stack height and roll spacing errors. This can be countered by setting the sine bar to produce the compliment of the desired angle (= 90° minus the desired angle), and then placing a square-all-over angle plate (or a toolmaker's block, or whatever type of fixture suits the job at hand) on its side beside the sine bar, make the set-up, and then roll the job and the holding fixture 90° to get the workpiece oriented the way we want it.

WHERE NOT TO USE A SINE PLATE

You've undoubtedly seen catalog pictures of sine plates complete with holes or T-slots and you've probably thought to yourself, "What a handy gadget that would be for my milling machine!"

I thought so too, but I wondered if such a device could be rigid enough to be used for milling operations. I asked the owner of a local tool and die shop about this. He said if he found one of his men so using a sine fixture, he'd fire the guy on the spot; six months of such use and the tool would be scrap.

In short, sine vises, sine plates, etc. are not for shaping, milling, etc. They are for set-up work, inspection, and surface grinding, where only light cuts are involved. Don't buy a high precision sine fixture for milling machine and shaper use.

Having said that, I will add that I have seen a simple home-made sine plate, whose maker wanted to cut some bevel gears. He put his dividing head on the sine plate, and did the job. No reason not to do this if the sine plate is "sacrificial" in nature - i.e. not a high class unit bought for a princely price, and dependent upon careful use for long, accurate service. Let's have a quick look at.....

A QUITE-GOOD-ENOUGH SINE PLATE

A shop-made sine plate can be made which will be found entirely serviceable, and useful for making angle set-ups which may be difficult or impossible with other available workholding equipment.

The "gage block stack", or packing, doesn't need to be Jo'blocks, accurate for size within millionths of an inch - the average hsm doesn't need that degree of accuracy, and doesn't have $150 to $500 or more to lay out for a set of toolroom grade Jo'blocks He can get by very nicely with a set of shop blocks, or even with various bits of stock size CRS, and shims of measured thickness. We've already looked at the errors, etc., of this whole business - no need of more words on that score here. Below are drawings for a simple sine plate, showing leading dimensions and enough info to enable anyone needing this type of fixture to hatch up one sized to suit his own equipment.

IDEA: A simple shop-made vice, such as the one shown elsewhere herein, could be designed as a detachable partner to this sine plate, and the two would at times make a very handy combo.

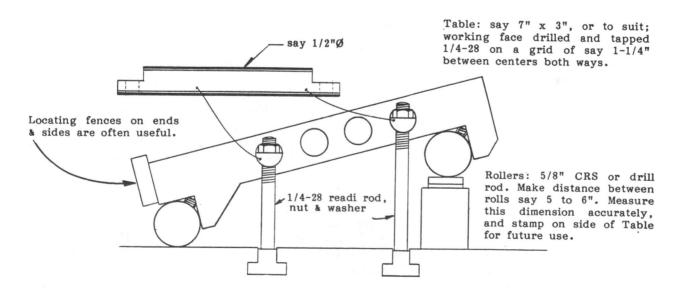

say 1/2"Ø

Table: say 7" x 3", or to suit; working face drilled and tapped 1/4-28 on a grid of say 1-1/4" between centers both ways.

Locating fences on ends & sides are often useful.

1/4-28 readi rod, nut & washer

Rollers: 5/8" CRS or drill rod. Make distance between rolls say 5 to 6". Measure this dimension accurately, and stamp on side of Table for future use.

79

SOME MILLING VICE ACCESSORY IDEAS

Most machinists have three hands, and are still not satisfied. If you are not so equipped, you will have noticed the lack, and the inconvenience often caused thereby. For example, you want to set up a job in the milling machine vice at some specified angle. You need to hold the protractor and the job - that accounts for two hands - and you will need to tighten the vice - that'll occupy your third hand - and then you may be seized with an uncontrollable urge to scratch your nose.

Having experienced all of the above and more, I am a sucker for things that will help out in such situations. Travers Tool (and others) offers a precision tilting V-block with a vernier protractor engraved on the side, and just about the right size to stick in my milling machine vice. Now here is something a guy can relate to! I ordered one, and I like it - you can set a job in the V-block, having previously tilted same to the desired angle and locked it there. Then, with this gadget and your job in the milling vice, you have two extra hands free to hold the pusher piece you may need to interpose between moving jaw and job, and to tighten the vice.

A QUICK DETACHABLE "SINE FIXTURE"
for your milling machine vise

Tim Smith sent me drawings of another good idea. One of the guys he works with has reamed a few holes in a plate which he can attach to the fixed jaw of his milling vise. Into these holes he can put pins which then provide a means of supporting a job at one or another of the several fixed angular settings provided by the hole pattern.

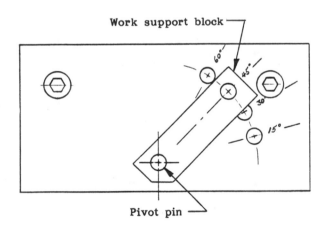

Work support block

Pivot pin

STARRETT "HOLD DOWNS"

Here's another workholding idea you may find useful, altho its use is not commonly shown in books of machine shop practice:

Starrett makes 3 sizes of "Hold Downs", which are used in pairs, most often in a milling vise, to force a workpiece - a piece of sheet brass, say, or a casting that's been cleaned up somewhat - flat down against the floor of the vise (or against a packing piece placed in the vise to protect the floor of the vise). "Hold Downs" can also be used with a set of toe-clamps or Mitee-Bites (Reid Tool; see Appendix), if you want to skip the milling vise and suck a job hard down onto the milling machine table directly.

The edges of these hardened "hold downs" are ground at an angle such that the pressure of the vise makes the edge in contact with the workpiece want to rotate downward. Thus the job is held securely, and all tendency for the job to rise is eliminated, while at the same time the entire top surface of the job is clear for machining. I have a pair, for which I foresee numerous uses in machining some of the sheet brass parts in the skeleton clock I plan to build when I get this book done!

MAKING A SET OF ANGLE BLOCKS

We have looked at gage blocks and sine bars; there is another way of making precise angle setups, and it is not beyond the reach of the hsm who might want it.

Sets of "angle blocks" are offered by Travers Tool and similar outfits, starting at about the $100 mark. These are sets of 10 hardened steel wedges, accurate to about ±20 seconds, in 1° steps from 1 to 5°, and thence in 5° steps to 30°. 1/4° and 1/2° blocks are offered as extras.

While hardly topping the "must have" list, some hsm's would probably find a set of these handy from time to time.

I think a fella could make a good set for the price of his time, and it'd be an interesting exercise. I would like to do this, perhaps for at least one or two useful angles.

If you want to make a set, you will need:
      a 5" or 10" sine bar,
      a set of shop blocks, or similar,
      a surface plate, and
      an angle plate.

Besides these and the techniques described elsewhere herein in the *American Machinist* article on making a flat master square, you require only time, patience, and some 1/8" or 1/4" hot rolled mild steel plate. This would be hacksawed roughly to size, milled nearly to finished specs, and then patiently worked down to finished form.

The type of material suggested is based on the *M.E.* article on "How To Make a Master Reference Square". Commercial angle blocks are 1/4" thick; thinner material might suit you better, depending on your own intended use.

Test each piece as it nears completion by setting it on the sine bar, which has been set for the angle desired, and placed up against an angle plate. Orient the workpiece so that the angles cancel each other out. Test by "indicating" the top edge of the angle block. If the edges are true, and the actual angle in the block is on spec, the indicator will read zero-zero from one end to the other.

It is unlikely that a hsm would require angle blocks more precise than say ±1 minute. If you were making a 15° angle block about 3" long, you would check it on the sine bar as described above. If the indicator reading is zero at one end, and ±0.001" at the other - again assuming the edge would test true against a straight edge - the error will be just about ±1 minute. If a fella wanted or needed better than that, I think he could produce it without exercising undue effort or patience.

ANGLE GAGE BLOCKS

"Angle gage blocks" provide a level of accuracy beyond that of the angle blocks referred to above. Like other gage blocks, they are made to exceedingly high levels of accuracy, in fact ±1 second and better.

Like myself, you might wonder how this extreme angular accuracy is achieved. I think it is correct to say that no home shop machinist can hope to produce such accuracy levels, nor does he need them*, but some info as to "how the factory does it" may be of interest. I asked the Starrett people if they had anything they could send me on the subject. Pretty soon along comes a nice 4-color booklet on the products of their Webber Gage Division, but it didn't tell anything about how they go about making the stuff. And fair enough - why should they give away details of their production methods?

                                                           *see note at top of next page

*Hi-precision angle gage blocks are important in military aircraft and rocket design, and in space guidance and tracking systems, where even extremely small angular errors are significant due to the speeds and distances involved.

Then a couple of weeks later I got a nice letter from George B. Webber, VP, Webber Gage Division. Mr. Webber sent me a copy of the eulogy given at his father's memorial service in 1984. (GBW's father was George D. Webber, founder of Webber Gage.) While I found it interesting, at first reading it didn't seem to contain any info on "How do dey git dem blocks so akurit?"

After a while it dawned on me that in fact Mr. Webber's eulogy actually did contain the key to the basis for the development of a set of Angle Gage Blocks, although nothing about the methods by which the blocks are produced on a commercial basis to the phenomenal level of accuracy they provide, beyond the fact that several Blocks of a given angle are ganged together and lapped to spec in a purpose built machine.

Webber's Angle Gage Blocks cover the range of 0 to 99° in steps of one second, with just 16 pieces. In the Tool Room grade, which has an accuracy level of ±1 second, they throw in a 6" parallel and a 6" knife edge straightedge. If you want real up-town accuracy, and do not lust for small angle capabilities, you can get their Laboratory Master set, which consists of 6 blocks covering 0 to 99° in 1° steps, with an accuracy of ±1/4 second!! Like I said, real up-town stuff.

These Angle Gage Blocks far exceed the accuracy needs of most of us, and at a cost approaching $2,500 for either set, they're completely out of reach of most of us, too. Of course, that they can be had at all is nice to know, if one has a real need for them, in which case the price will not be a problem, either.

The development history of these blocks bears mention also. The British think that they developed them, but... on December 30, 1935, one Nikola Trbojevich filed for a US patent on an invention entitled "Gages for Measuring Angles", and on October 25, 1938, was granted Patent No. 2,134,062.

Now Trbojevich had the patent, but not the knowhow to make the actual item. Webber Gage recognized the value of Trbojevich's idea, and obtained exclusive rights to this landmark invention in late 1942. Three years elapsed before the Webber people finished their first set of Angle Gage Blocks, and sent them to the US National Bureau of Standards for examination.

Let's consider how one would go about originating, from more or less nothing, a set of Angle Gage Blocks, following the info found in Mr. Webber's eulogy.

If one had a perfectly frictionless precision balance, and a piece of steel known to weigh exactly one pound, he could, with the aid of said balance develop a whole set of weights. He would first make two new blocks of steel, each exactly equal to the other in weight, and together weighing exactly the same as the original one pound block. The process could be repeated to finer and finer levels of division, "to taste".

By the same token, if one had a rectangular block known to be exactly 90° on every corner*, he could set about making two 45° blocks - they would need to be worked upon until they were exactly alike, and together exactly square with the original 90° rectangle. Then, three 15° blocks, all identical, and together exactly matching either 45° block, and so on until one had worked up a complete set of Angle Gage Blocks, plus a few left over pieces besides.

And that, in a nut shell, is how Angle Gage Blocks were originated - patience, and more patience!

*We have seen elsewhere herein that a true right angle can be derived from nothing, if one has a plane surface (= 180°) to work from.

Choosing and using
A SENSITIVE DIAL INDICATOR

If you are interested in originating reference-grade squares, toolroom-grade angle blocks, space blocks and similar (relatively) high-standard work, including spot grinding and lapping, you will need a sensitive dial indicator (DI). It need not have a great range - 10 or 20 thou would be plenty - but it must be very sensitive.

A couple of years back I lucked onto a Starrett #656-517J, new in the box, at a price that brought my wallet out with my thumbnail smokin'! The 656 series indicators have a 3-5/8"Ø dial. This particular one has a range of 0.400" x 0.000,1", and jewelled bearings. I think a Starrett #25-116J (2-1/4"Ø dial) with 0.015" range x 0.000,1", or some other make of similar specs, would be a better choice, as it would be lighter and therefore easier to support on a DI base, etc.

Another good choice would be the Starrett "Last Word" #711D-10 indicator (0.024" x 0.000,1"), which runs to about the same kind of money. This type is probably ideal for the purposes of the amateur gagemaker: it is light, compact, and very sensitive. Its lightness permits it to be mounted on the spindle of even a toolmaker's surface gage - a very handy arrangement. Comparable indicators by Mitutoyo or other reputable makers would serve as well.

To be truly useful for this type of work, a DI must be able to reveal a drop of the plunger of say 0.000,1". Many a DI will happily report a drop of say 0.001", but not one so small as we would desire, although it may be able to report a **rise** of one "tenth". If testing a work piece for parallelism by passing it under a DI on a surface plate, do so by moving the work under the DI in both directions, and in each case by running off the end of the work.

If the work appears to be "going up hill" one way, but not down hill the other way, your DI is faulty. If it be sensitive enough to show the error involved, the "going uphill" reading should then be taken as correct.

Another test to try is passing the work under the DI backwards and forwards - this may give different readings. If so, there is probably some slop in the indicator's plunger bearings, which is not likely fixable. So long as the variation in the two readings is not large, you are still ok - just make sure you use the DI in a consistent manner.

Naturally the DI needs to be mounted on a very rigid stand, having also fine adjustment capabilities.

AN ULTRA SENSITIVE DIAL INDICATOR BASE

The following drawing shows an ultra sensitive dial indicator base. I think it would be fairly easy to make up something like this, and that it'd be well suited for use with a tenths indicator:

Very fine adjustments would be had by sideways screw pressure on the internal cantilevered beam, which would bend the tube even less, and the DI would move very little, but enough. This idea was developed by England's National Physical Laboratory in the 1960's.

Two features which bear comment are the base of the Tube, and the bevelled edge of the Base casting.

In the illustration I saw, the steel tube was shown thickened at the bottom end. I suspect that in making the prototype, the tube used was fairly thick-walled stuff, and that the O.D. was ground to a high finish and uniform dia. over the working length. This would give both easy movement for coarse adjustment, plus a thinner wall, for easier bending. At the same time, leaving the base of the tube full thickness would allow a heavier press fit between the tube and the Base casting. The tube was probably pressed in from the underside of the Base - certainly this is how I would do it. (Or Loctite it in place.)

The bevel on the Base's working face would help protect it from edge injuries.

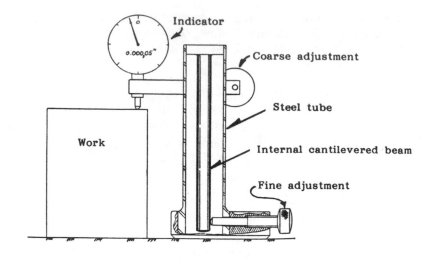

GA drwg of an ultra sensitive Dial Indicator Stand developed by Britain's National Physical Laboratory in the early '60's.

A DIAL INDICATOR CLAMP

Al Zueff is a tug boat captain here on the West Coast, and has held his Master's papers for something like 38 years. His enthusiasm for the machinist's trade as a hobby becomes evident at short acquaintance. He showed me the following idea - an extension arm on which to mount an indicator, said arm then being stuck into the final clamping sleeve of one's magnetic base. I think it's a good idea - the only reason I have not yet made two of them, one a bent, or goosenecked version, is this book what I am busy writing.

Al's surname is pronounced "zoo-eff", but when I get around to making one of these gadgets for myself, I'm going to call it a Zuff-stick.

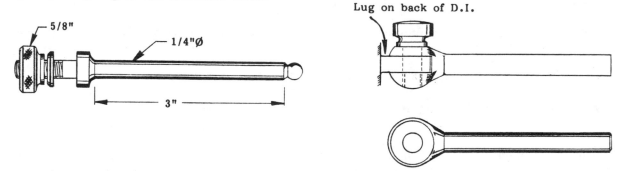

HOW TO MAKE DOVETAIL DIAL INDICATOR CLAMPS

Some lever-type dial indicators have a dovetail incorporated into the indicator body - this mates with holders of various form, the business ends of which need to look about like as at Fig. 1.

I figured I'd like to be able to make my own adaptors, so I got a small dovetail milling cutter from Brownells, and proceeded to break it on the 2nd of two adaptors that I had ready and waiting for the cutter's arrival in the mail. I finished the second one more by crudities than finesse, but the cutter was done for anyway, so I had nothing to lose.

But that's not the way I like to do things, so I decided I had to find some other means of cutting the dovetail, and find it I did.

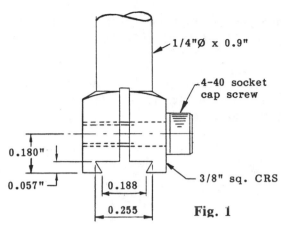

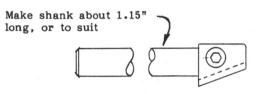

1/4"Ø x 0.9"

4-40 socket cap screw

0.180"

0.057"

0.188

0.255

3/8" sq. CRS

Fig. 1

Make shank about 1.15" long, or to suit

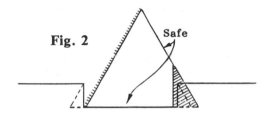

Fig. 2

Safe

Grind off hatched portion of file so file can enter starting notch.

Brownells sells what they call a "sight base file". It is a parallel triangular file with two safe sides, and costs about $9. I ordered one, and found it needed a little modification, per Fig. 2, after which I had a durable means of making all the dovetail adaptors I could ever want.

What to do is make the adaptor blank, and run a 3/16" slot drill through to a depth of 0.057" to form the basic channel, then use the sight base file to cut the dovetail by hand. Sounds not very easy? In fact it is not difficult: the safe edges let you cut exactly where you want, and not cut at all where you do not want.

The dovetail is a clamp, not a working set of dovetail ways, so perfection is not required - just excellence: the filed-out dovetail need only pass the male dovetail without binding. Make it as good a fit as you can, and it'll serve quite nicely.

I have shown a basic type of adaptor here. Make yours to suit a particular need, or several when you are in the mood, because if they will permit you to mount the indicator at all, the day will surely come when you will have a use for every one of them. Such things are good projects when you don't have bigger jobs to tackle, or don't feel like doing them.

A REQUEST FOR IDEAS

Does anyone have any ideas for a design for a master vernier protractor for use on setup work on the surface plate and on the milling machine table?

A vernier protractor is an expensive item to buy, yet every machinist needs one. I have both a vernier protractor and a rectangular head protractor/depth gage, both by Starrett. Both see frequent use. On occasion I have stuck the vernier protractor to a magnetic base, in order to have one less thing to hang on to while making a setup of one sort or another. I've often thought that a useful thing to have would be a master vernier protractor which could sit on edge on the surface plate, or on the milling machine table.

"Lautard's OCTOPUS" is a step in this direction, in that it has a vernier protractor incorporated into its design, but what I am thinking about here is more of a one-purpose tool. I have a few ideas, but am not very far along with it.

In the course of my shop activities I have had occasion to produce a number of dials and 0-360° protractor scales, and a couple of 5 minute verniers to use with one of the latter. Some are cut radially, some axially, and some are cut into a coned face. My methods are detailed in the first chapter of my book *Strike While the Iron is Hot*, which is excerpted in **TMBR**. Those of you who have either of those books have therefore read most of my ideas on the subject, and if you also have my working drawings & instructions for making **Lautard's OCTOPUS**, you've read some more info there.

Thus I know that with easily made cutters one can readily engrave a 0-360° protractor face that is as handsome and accurate as any protractor that ever came out of a factory.

I think any hsm could make himself a first class vernier protractor, if he had a good design to work from. Like I said, I have some ideas, but am not very far along with them. Therefore I put this question to all who read this book:

Have you ever seen, or have you any ideas for, a vernier protractor which would stand on edge, and which could be readily made from say a 3.5" square of 1" HR MS plate, plus a piece of ground flat stock, the rest probably coming from the scrap box?

If you have any ideas along these lines, I would be very pleased to hear from you. What I have in mind is to work up a good design, make a couple of prototypes, then work up a set of drawings from which anyone else could make a similar animal.

A SIMPLE SMALL VISE

A small vise can be built up from CRS along the lines shown below. If you want greater precision, make from hot rolled steel, machine all over, rough grind, caseharden and then either surface grind, or spot grind and lap. The drawing shows the general idea, which is adaptable to many uses, from a drill press vise up or down to a tilting angleplate accessory, or a miniature vise for some special application all your own.

Dowel and silversolder together the body of the vise, or assemble with 1/4-NC socket cap screws. Use a pair of 5/16" NC socket caps for the pressure screws - bevel the ends, or brass tip them by drilling into the ends and fitting separate brass tips, or build up a brass tip by brazing and then turning down to civilized contours.

Attach to any T-slotted machine tool table with toe clamps - see drawing. Various "pushers" can be made to suit the general run of work, and for special jobs as they arise. One useful type would be one having horizontal and vertical V-grooves.

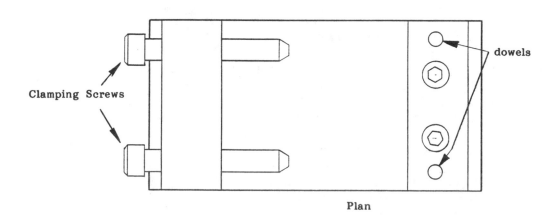

Plan

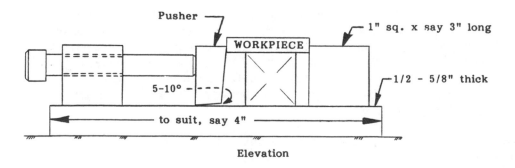

Elevation

CUTTING MULTIPLE START THREADS
by Mac Mackintosh

(Mac wrote this item specifically for **The Machinist's Second Bedside Reader**. He has also published details of this idea in *The Journal of the Society of Model and Experimental Engineers*, and in *Model Engineer Magazine* for Nov 16, 1973, p.1130.)

Multiple start threads are sometimes required for quick-release bolts. More often, they are required for translating threads such as are used in worm gears. Sometimes it is possible to juggle with the threading dial to get the necessary starts but often it is not possible to fit the dial indications to the pitch of the thread – in this case what is often done is to set the topslide parallel with the lathe axis, cut the first thread, and then set the topslide forward by 1/2, 1/3, or 1/4 of the selected pitch, depending on whether a 2-, 3-, or 4-start thread is being cut.

With the topslide set parallel for this purpose, it is impossible to use an angular setting of the cutting tool, and the tool must then be advanced into the workpiece at right angles. With a V-tool, the resulting finish is usually rough, owing to interference in swarf clearance due to the fact that two faces of the tool are cutting simultaneously.

Most multiple start threads are required for translating threads; the most efficient form for these is the Acme thread. In the case of an Acme threading tool, the swarf clearance difficulty is further compounded: owing to the flat at the tip of the tool, it has to cut on three surfaces simultaneously. It is very difficult to get a good finish if an Acme tool is advanced directly into the workpiece. If the topslide could be slewed to 14.5°, the tool would cut on two faces only, and the swarf clearance would be greatly improved, since the amount of stock being removed by the land at the end of the tool is small.

Some years ago I made up a fitting that would allow multiple start thread cutting with an angular setting of the topslide. For want of a better term, I call it an **Indexing Faceplate**. It allows the workpiece to be indexed by a precise amount and held there; as the workpiece is independent of the topslide, the latter can be set at an angle for advance into the workpiece.

It consists of four pieces (two assemblies) as shown in Fig. 1. Dimensions shown are to suit my lathe, which has a 2-1/4 x 8tpi thread on the nose. You will modify the dimensions to fit your lathe.

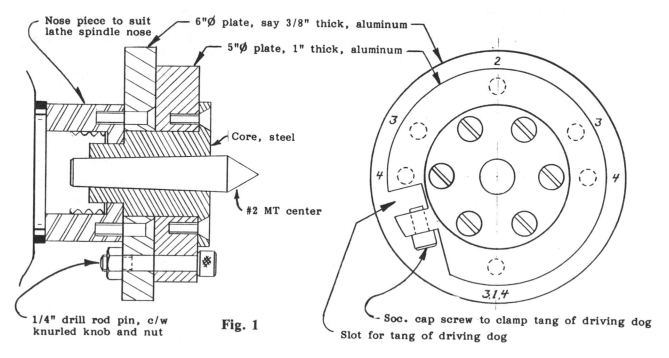

Nose piece to suit lathe spindle nose — 6"Ø plate, say 3/8" thick, aluminum

5"Ø plate, 1" thick, aluminum

Core, steel

#2 MT center

1/4" drill rod pin, c/w knurled knob and nut

Fig. 1

Soc. cap screw to clamp tang of driving dog

Slot for tang of driving dog

An Indexing Faceplate, for cutting multiple start threads

The first assembly consists of a steel nosepiece to fit the lathe, to which is screwed a plate - in my case this plate is aluminum, 6"Ø, and 1/2" thick. The second assembly consists of a steel core to which is screwed a second plate of aluminum (mine is 5"Ø, 1" thick); a hole is bored in the 6"Ø plate in the first assembly to a close rotational fit with the diameter of the steel core of the second assembly.

A slot is cut in the 5"Ø plate and a screw fitted so that a lathe dog can be held rigidly. A hole is drilled and reamed opposite the slot at a suitable radius for a 1/4" hole. (NOTE: this hole in the 5"Ø plate is not shown in the end view of this device. GBL)

Using this hole as a guide, six 1/4" holes are drilled and reamed in the 6" dia. plate as shown in the drawing. A 1/4" screwed pin is made for these holes from drill rod and is fitted with a knurled knob and nut. The holes are stamped to indicate which holes to use for cutting 1-, 2-, 3- and 4-start threads.

The toolholder shown in Fig. 2 will allow a single-start tool to be tilted to the correct helix angle of a particular multiple-start thread. As it is designed to be clamped directly to the topslide, the square hole in the inner sleeve should be located so that the cutting edge of the tool comes at lathe center height.

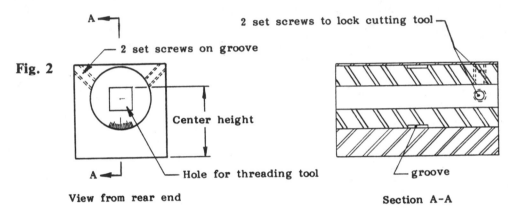

Fig. 2

View from rear end

Section A-A

How to make a "square hole sleeve*":

Mill a channel down the length of a piece of stress relieved square CRS of suitable size, and twice as long as the desired Sleeve's length. Cut in half. Silver solder the 2 pieces together, with the channels face to face so they form a square hole of the desired size. Using your 4-jaw chuck, accurately center, and then center drill the ends of a piece of square stock to fit the hole, and slide the unfinished Sleeve onto it. Drill and tap the Sleeve for the two set screws that will later lock the toolbit into the Sleeve, fit same, and then use them (or possibly epoxy) to lock the Sleeve and the square mandrel together. You can now readily turn the Square Hole Sleeve to a good fit in the reamed or bored hole you will have already made in the main Square Toolholder.

Turn the groove at center of the Sleeve where the other 2 set screws will contact the Sleeve. Dechuck, remove the square mandrel, chuck the Sleeve in your 3-jaw, and face the ends. Engrave the grads on the end that will be facing you when the device is in use.

*An item by this same name can also be bought, but is not suitable for this application, as the OD is not much over the across corners dimension for the square hole (which is no doubt broached) therein, which leaves insufficient wall thickness for the dual setscrews in the sleeve wall. GBL

Since this accessory is primarily intended for Acme threads, a further problem arises in grinding cutting tools. It is necessary to have a separate tool for each size of Acme thread, due to the difference in width of the flat at the end of the tool. I have made myself a set of tools for 16-, 14-, 12-, 10 and 8-t.p.i. Acme threads, right-handed at one end and left-handed at the other, but these are for single-start threads. With single-start threads the normal cutting angles can be adhered to, but with a 2-start thread the side rake becomes exaggerated, and with 3- and 4-start threads, the side rake becomes so exaggerated that the tool becomes so weak that it may break as soon as it is used. Quite apart from difficulties in grinding such a tool, lathe tool blanks are expensive and if one were to make a set to cover single and all the other starts, there would be some 20 tools in the set.

If the Acme tool is tilted, we can get away with a slightly modified set of single-start tools for the multiple threads. Machinery's Handbook gives 10-15 degrees for side rake in alloy steel. If we grind 15 degrees side rake into our tools, this should take care of the extra "curl" in a multiple-start as compared with a single-start thread, provided that the single-start tool is tilted to the correct helix angle for the thread being cut.

———————————∿∿∿———————————

A USEFUL ACCESSORY FOR CUTTING TAPERS IN THE LATHE
by Mac Mackintosh,
(with a few thoughts tossed in by the other guy)

(Mac wrote this item specifically for **TMBR#2**. He originally published this idea in *The Journal of the Society of Model and Experimental Engineers*. He had an article on the same subject in *Model Engineer Magazine* for March 7/69, page 250.)

The Adjustable Center shown here has been a tremendous time-saver in making taper shanked tooling for my lathe.

Quite apart from the cost of tools nowadays, it is often convenient to be able to make one's own tools because frequently the exact item required is not available from suppliers. Often it is necessary to make a special tool that will be operated from the tailstock; this generally necessitates cutting a taper to fit the taper in the tailstock barrel.

(Blank end taper shank arbors - and that is their correct name - are a standard product of the drill making trade - if you look at several taper shank drills you will see why: in many cases the drill is silver-soldered into the shank. Thus it is surprising that blank end taper shank arbors are not more readily available. Most good tool suppliers, even if they don't stock them, will know of them, and can get them for you, but I have encountered not a few tool merchants who will give you a blank stare when you ask for a blank end taper shank arbor. When you do find them, the price may surprise you - one would think they would be priced about like a drill chuck arbor, say 3 or 4 bucks, but they are more likely to run about $10 to $15 or more. GBL)

There are a couple of outs. The first, if you can use a drill chuck arbor, is to buy exactly that, and turn down the end to suit your purposes.

The second alternative is to make your own taper shanks, right in your own shop. If you only want one taper shank, buying a drill chuck arbor may make more sense, particularly if you put much value on your time, but if you will turn out a dozen while you are at it, making them yourself becomes quite practical - so long as you have use for that many!

The simple way to do it is to set your taper turning attachment to the correct taper, and then whittle up a bunch of them. Tap the butt (small) end of each shank before cutting it off of the parent material - you will likely want to use a drawbar with at least some of them, eventually, so best to make provision for it now. Lacking a taper attachment, slew your top slide over and use it to get the desired taper. Either way, getting the lathe to cut the correct taper will take some trial and error, so once you get everything set right, make several.

(My experience of turning fitting tapers is limited primarily to a few one-off jobs, plus having done one bunch of #2M.T.'s with the topslide-slewed-over method; it worked fine for me. But read what Mac says, in the next couple of paragraphs. GBL)

Tapers are divided into two categories: self-releasing and self-holding. (Which also see, after this section. GBL) Self-releasing tapers (i.e. relatively steep tapers) are best cut by setting the topslide at the correct angle (which, as we have said, is a matter of trial and error) but self-holding tapers are a different matter - there are two methods usually used for cutting these.

Taper turning attachments, supplied by lathe manufacturers as an accessory, are expensive and are rarely satisfactory in a light lathe, owing to the fact that the pressure of the cutting tool tends to swing the whole lathe carriage a little, with the result that the true taper does not begin until the carriage has travelled maybe half an inch. This means that the taper being cut cannot be checked against a female taper until the faulty section has been removed.

> NOTE: I would think that this could be corrected by having the saddle gibs properly adjusted. However, I can picture what Mac is saying here, and his experience of taper turning attachments must exceed mine; I have a taper turning attachment for my Super 7, but have not yet had reason to use it. GBL

The other method is to set over the tailstock on its base by means of its adjusting screws, and then turn the blank between centers. This is an entirely satisfactory procedure so far as accuracy is concerned, but it is time consuming to get the setting exact even if there is a dummy taper available to set against a dial indicator. And there is the equally unwelcome and exacting task of setting the tailstock back to dead center after the taper has been finished. (NOTE: See the tip at the end of this piece for an aid to re-centering the tailstock. GBL)

What would be nice would be to be able, at a moment's notice, more or less, to make a single #2M.T. shank in a piece of 3/4"Ø mild steel costing only a matter of a few cents. The Adjustable Center accessory described below gives you the ability to do just that. If it is well & carefully made, it will do it's job excellently; if carelessly made, it will be useless.

Years ago, I read an article in *M.E.* #3303 by "Duplex" in which a "set-over" tailstock center was described. This device, while it did not eliminate the fussy job of correct setting for the taper, did at least avoid the need to upset the alignment of the lathe's tailstock to do it. It occurred to me that a small modification would allow quick setting for a number of different tapers whenever one or another was required.

The center which carries the work is mounted on a Slide at right angles to the tailstock barrel; Duplex used screw threads for anchoring the center and shank into the slides, but I discarded these as lacking in accuracy and decided to use press fits instead. Fig. 1 shows Duplex's arrangement; Fig. 2 shows my modification.

To make one of these useful accessories, the following materials are required: a 3.5" length of 3/4"Ø mild steel (or a purchased #2MT drill chuck arbor), two 3" lengths of 1" x 3/8" CRS, a piece of 1/2" or 3/4" drill rod, a 3/16"Ø x 1" dowel pin and two 1/4-28 x 3/4" socket cap screws.

Make the Slide first. In one of the rectangular pieces, mill a groove 1/2" wide x 1/16" deep; two 1/4" slots are then milled 5/8" in from each end as shown. (Note: It would do no harm to stress relieve the CRS used for both parts of the Slide before starting to machine it. For info on stress relieving, see **TMBR**, page 25. GBL)

The male part of the Slide is then milled to a nice sliding fit with the female part, and two 1/4-28 holes drilled and tapped as shown.

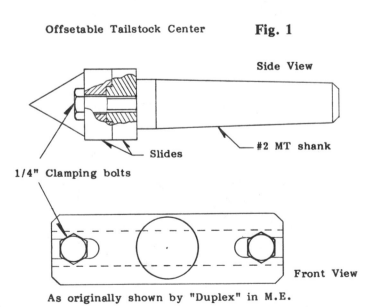

Offsetable Tailstock Center   **Fig. 1**

Side View

Slides

#2 MT shank

1/4" Clamping bolts

Front View

As originally shown by "Duplex" in M.E.

90

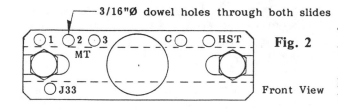

3/16"∅ dowel holes through both slides

Fig. 2

Front View

As modified by "Mac" Mackintosh

The two parts of the Slide are then bolted together and the center of the Slide located; a 1/2" hole is drilled and reamed through both parts of the Slide while they are bolted together, thus ensuring that the holes in the two parts are exactly in line. The two parts of the Slide can then be separated again.

If you have bought a drill chuck arbor, you will not have the job of cutting one last taper by the old method of setting over the tailstock. In any case, having made or bought a taper shank, its "business end" is turned to a 1/2"∅ press fit 3/8" long in the male part of the Slide.

Once the taper shank is done, it is carefully pressed (or shrink fitted) into the male part of the Slide. Allow about 1.5 thou interference for this fit. If doing a shrink fit, make sure you have a good plan of action for getting the 1/2"∅ end of the taper shank straight into its hole - if you "miss" in any manner, you will not get the two pieces apart for a second try. If doing a press fit, put some white lead on the 1/2"∅ stud before pressing the two pieces together.

The piece of drill rod is turned to a 60° center, and polished; behind this turn to a 1/2" press fit for 5/16"; thus, with almost no effort, you can ensure that the 60° center is concentric with its 1/2"∅ shank for the press fit. The Center is hardened and pressed into the female part of the Slide. (Interjection: If you don't want to do press fits, think about Loctite. GBL)

I decided that I would arrange my Adjustable Center for quick setting to cut Nos. 1, 2 and 3 Morse tapers, for dead center turning also if wanted, and to cut a No. 5 Morse taper to suit my headstock taper. (That's the hole marked "HST" in Fig. 2.)

I have reamers to produce Nos. 1, 2 and 3 Morse taper sockets, and decided to use these as models. I tackled the No. 2 MT first. I put a center in the headstock and the new Adjustable Center in the tailstock, and set up the taper reamer between them. The holding bolts for the Adjustable Center were half tightened and the Slide set over with light taps of a soft-faced hammer until a dial indicator traversed from end to end of the cutting edges gave a zero reading. The holding bolts were then fully tightened, a blank was substituted for the reamer, and the taper cut using a fine power feed.

The blank was then taken out and checked against the tailstock socket - further slight adjustments had to be made until the taper was as good as I could make it. At that set-over setting of the Adjustable Center, a 3/16" hole was drilled and reamed through both parts of the Slide for the dowel pin.

A similar procedure was followed for the Nos. 1, 3 and 5 Morse tapers. Dowel pin holes were also provided for dead center. I have since put in a pair of holes for the No. 33 Jacobs taper, and I still have room for five more tapers (that is to say, dowel pin holes at other taper settings) if needed.

In use, this Adjustable Center tool can be set up just about as quickly as putting in a fixed center; its accuracy is such that a few minutes work with a fine file and hand stone will produce as good a taper shank as a professionally ground one.

This "finishing up" work will be needed because the lathe toolbit will leave a finish which, for a taper shank, one would consider rather rough. It is therefore necessary to do some finishing work as noted above - 5 or 10 minutes work only - any small discrepancies being taken out during the process. Check the job against a taper socket gage (see below) using layout blue, or 3 chalk marks drawn down the length of the taper and spaced about every 120° around it, as you work.

When plugging the Adjustable Center into the tailstock for use, I find it is entirely satisfactory to set it horizontal by eye. While making the attachment in the first place, I set it horizontal with an indicator; one wants to eliminate all sources of error when making a tool, even though the same degree of care may not be required when using it.

Note: When using the Adjustable Center, or the set-over tailstock method in general, the length of the blank being worked on, (and, to a lesser extent, the depth of the center holes) will have an effect on the taper produced. The taper shank reamers were chosen for models because they are longer than the overall length of most parts requiring a taper at one end.

The length of blank to use for each taper desired should be carefully recorded for future reference.

If the item you are making calls for a taper shank of a lesser overall length, cut the blank to the standard length, and cut to the desired OAL after the taper has been cut. My procedure is to measure the length of the blank with a vernier caliper, and put a #0 center drill to full depth in each end. That is close enough.

> (TIP: This may be as good as any place to interject the following point, which is not written in blood & stone, but probably a good basic rule: The correct depth to drill a center hole is normally about 2/3rds the length of the conical part of the center drill. If the workpiece is heavy, one might enter the drill a little deeper. GBL)

To finish off, here are some tips you may like:

> 1. If you want a master gage to test your taper shanks against, buy a hardened and ground "drill sleeve" having the desired inside taper. A #2/#3 costs about $6, and would be cheap at twice that. With this at hand, there is no need to remove whatever tooling is mounted in the tailstock at the time, in order to use the t/s socket as a gage.

> (If you also want to have a ready means of adding a tapped hole to the end of shop made (and bought) taper shanks, buy instead what is known as a "solid socket". These have a parallel OD, and cost maybe $10. When you want to put a tapped hole in the end of a taper shank, bang your taper shank into the solid socket, and chuck the latter with the small end of the shank looking sheepishly at the tailstock. The rest is easy.)

> 2. If you just don't want to fuss around a whole lot, to get the taper perfect, turn down the blanks to pretty near done, and then take them to a tool & cutter grinding shop to clean up and bring them dead on spec. Obviously, if you do this, you will want to provide a good center hole at both ends for the t&cg shop's convenience. Also, for the same reason that you would make several at one go if doing them entirely in your own shop, you would do well to take several rather than just one to be finish ground.

> 3. A safety razor blade nipped between a good pair of centers will allow you to see quite readily whether the tailstock has been brought back into line with the headstock - if it angles even a little bit one way or the other in plan view, adjust some more - things aren't what they should be.

How to design
SELF HOLDING AND SELF RELEASING TAPERS
(plus some info on friction
of various metal/metal combinations)

Morse Tapers are "slow" tapers which are "self holding" - once slid home in a clean Morse taper socket, a mating taper shank will stay put (within limits) until bunted loose one way or another. A "fast" or steep taper will be "self releasing" - the term is self explanatory.

But when does a taper change from self holding to self-releasing?

For example, if you were designing a collet chuck with a non-standard taper, you would want to know what taper to use to ensure that the collet would not "freeze" in the collet closer. What to do?

Answer: The angle with centerline must be large enough that its tangent exceeds the coefficient of friction between the materials from which the mating parts are made.

Tim Smith (whose name comes up elsewhere herein, and who is a tool & die designer in his own right) sent me a clipping headed "From The Checker's Desk", by Clement F. Brown - no indication as to when or where published. The problem there cited is that of a collet-like clamping element employing a 7.5° cone angle; this was found to freeze in the closed position, i.e. it would not self-release.

"The Checker's" advice was to increase the angle to 10°. Now the coefficient of friction for steel-on-steel, un-oiled, is 0.1490; and tan 10° = 0.1763°, while tan 7.5° = 0.1317;

Now, lest thee get thyself into a great sweat and other difficulties through not paying attention, let me give thee a warning: if thee look up the coefficient of friction in Machinery's Handbook (page 553 in the 20th Edition), there thee will find the coefficient of friction for un-oiled steel-on-steel is 0.8 - which is a mighty big jump from 0.1490!

If you then blindly use the number 0.8 from Machinery's Handbook, and your application involves steel on steel, you would find the angle whose tangent is 0.8 (= about 38.5°) and you would probably think it prudent to use an angle of 40°. (Or to stop reading Lautard's stuff permanently!)

But if you look at a drawing of a 5C collet, the angle is spec'd as 10°. In the case of an R8 collet, the drawing I have does not give this info, but near as I can measure it, it is 8.5°, for which the tan is 0.149451, or just about exactly the same as "The Checker's" number of 0.1490.

In my experience, an R8 collet usually takes a light tap to get it out of the mill's spindle nose, and a 5C collet comes loose when you slack the retaining nut, but this might be due to minor differences in the finishes of the two respective collet sockets of which I have experience, rather than design per se. Therefore, the haunting question is, "Whose numbers are right?"

I phoned the Library, and in short order the librarian found a table in *Kent's Mechanical Engineering Handbook*, Volume 2, and read me the same number as "The Checker" gave, i.e. 0.149. I therefore feel reasonably comfortable (and useful) giving below a few numbers based on info in said table.

(A) For un-lubricated metal/metal contact, at rest:

steel/steel	0.149
cast iron/steel	0.155

(B) For un-lubricated metal/metal contact, in motion:

brass/brass	0.173
brass/cast iron	0.188
bronze/bronze	0.199
bronze/cast iron	0.213
cast iron/c.i.	0.130-0.140
steel/bronze	0.152
steel/c.i.	0.204

(C) For lubricated metal/metal contact, in motion:

brass/c.i.	0.111
bronze/bronze	0.133
bronze/c.i.	0.098
c.i./bronze	0.132-0.169
c.i./c.i.	0.130-0.140
steel/bronze	0.0004-0.0062
(various values, bath lubed)*	
steel/c.i.	0.108

*this would likely be for a shaft running in a bush type bearing, on an oil film. Where an oil film is maintained, friction is independent of the materials involved, but rather, varies with the viscosity of the lubricant being used.

BETWEEN-CENTERS BORING BARS

A between-centers boring bar effectively converts the lathe into a horizontal boring mill, and is thus a very useful lathe accessory. A rough hole - e.g. a cored hole in a casting - can be enlarged by such a tool to a desired size, and to a roundness and overall accuracy probably more easily achieved thus than with any other approach practical on the lathe, short of lapping.

Furthermore, the concept has wider possibilities, taken in somewhat different directions. We will look at these off-shoots in a minute, but first let's just consider the basic idea, as in Fig. 1.

Fig. 1

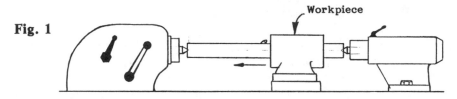

Between-centers boring bar in action

The bar should be made of a suitable size for the jobs expected, or for the job at hand. Its diameter must be somewhat less than the hole it is put into, but the fatter it is for a given length, the less spring it will have. Length must be such as to accommodate the job and not a lot more - again to avoid undue spring.

Counterbore the center holes at both ends to protect them from damage, as at Fig. 2. Put the hole for the cutter through the bar at an angle of about 45° to the bar's long axis, rather than straight across at 90°. The latter arrangement would give you a hole at the back (opposite) side of the bar, which is undesirable, as it prevents ready measurement of cutter protrusion, hence prediction of hole size.

Fig. 2

Fig. 3

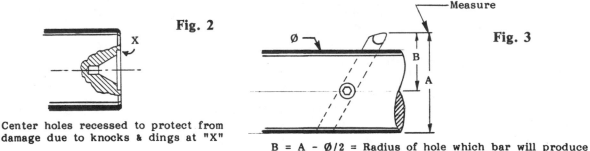

Center holes recessed to protect from damage due to knocks & dings at "X"

B = A - Ø/2 = Radius of hole which bar will produce with cutter at this setting

Fig. 4

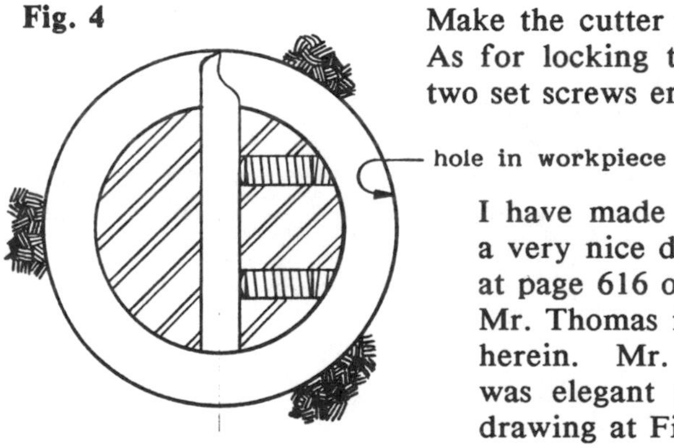

Make the cutter from HSS drill blank material, say 3/16 or 1/4"Ø. As for locking the cutter in the bar, a simple solution is one or two set screws entering from the side, as in Figure 4 at left.

hole in workpiece

I have made two between-centers boring bars, these following a very nice design by George H. Thomas, which was published at page 616 of the June 3, 1977 issue of *M.E.* I am indebted to Mr. Thomas for permission to present my version of his design herein. Mr. Thomas' means of locking the cutter in the bar was elegant and interesting - details are shown in the scale drawing at Fig. 5.

Making a between-centers boring bar of this sort is an interesting exercise, as it calls for making (and heat treating) a tapered shank locking screw, and an axial push rod.

If you want to make one like this, make an accurate scale drawing, say 3X full size, and work up the dimensions for yourself. (No sense me doing it, as you would just want one of a different size.)

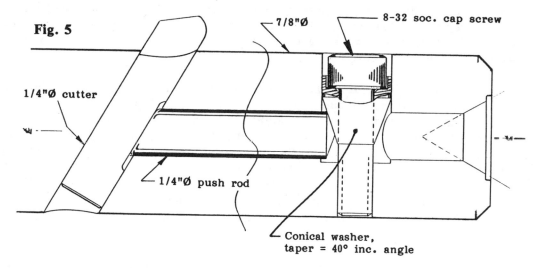

Fig. 5

7/8"Ø

8-32 soc. cap screw

1/4"Ø cutter

1/4"Ø push rod

Conical washer,
taper = 40° inc. angle

I went so far as to swage a hex socket in the heads of the locking screws in the two I made, rather than using a slotted head, as GHT showed. For each size, this required the making of a hex punch, with a slight degree of back taper for relief. A hole was drilled in the head of the screw blank just slightly larger than the A/F dimension of the hex to be swaged. I made a holder to drop the screw into, plus a sleeve to follow the screw, said sleeve being drilled to just nicely take the 1/4" drill rod from which the punch was made. This tooling kept the punch aligned with the pre-swage hole in the head of the screw blank. The swaging operation was done cold, pressure being applied with the bench vice.

If I were making another, I would probably just tap the end of the screw and Loctite a socket set screw in the hole - about as good, and less work. Easier yet: use a standard socket head cap screw, and a conical washer, as shown on Fig. 5.

The axial hole is gun drilled with the aid of a steady rest (unless your lathe spindle bore will accommodate the bar; lucky you, if it will!) The cross hole is drilled, tapped, and counterbored, the lock screw and push rod made, and the parts assembled. Next, set up the job at some angle near 45° in the mill, or at center height on the lathe cross slide, and cross drill/ream for the cutter at a location carefully figured to also cut the end of the push rod for its later service against the cutter. I find there is a flat ground on my cutter, which means that the push rod enters the bar a little further than its position when the cutter hole was reamed. To compensate for this, the push rod should be made long enough initially; the lock screw will be turned in deeper in use than during the reaming of the cutter hole.

TIP: To produce well centered holes in the ends of the bar, face and bevel both ends of the bar material, probably with the aid of a steady rest. Locate the center of each end as closely as possible with a center square. Or, blue both ends and set the workpiece on matched V-blocks on your surface plate, and proceed to locate the center of each end as closely as possible with a surface gage or height gage. Very carefully make an initial "center" with a scriber and the aid of a magnifying glass. When satisfied, centerpunch lightly. Set the bar between a pair of nice sharp centers in your lathe (Make sure it doesn't drop out from between them!) and check it with an indicator.

If not close enough, carefully note which way the centerpunch mark needs to be moved, take the bar out of the lathe, and apply the centerpunch at an angle to move the center over. With care, you can get the bar to run true within a thou or 2 on its centerpunch marks. When it is thus, deepen the centerpunch marks, and check again.

(If you do some thinking about it, you will see that the center holes in a between-centers boring bar could be off by 20 thou and make no difference, but we are trying for a high class job here, and the technique is useful anyway.)

Grab one end of the bar in the 3-jaw, support the other in the fixed steady, and drill the center hole. Reverse and repeat. Then recess the ends to provide built-in protection, as earlier mentioned - see Fig. 2. Once the centers are in, proceed to drill the push rod hole, and do the other operations to complete the boring bar as outlined above.

TIP: It sometimes happens that one needs to drill a job from both ends. The boring bar is not a good example - because we can cope with it via the chuck and a steady rest, and we only want to *center drill* that job - but I doubt this idea will fit in anywhere else any better, so I'll mention it here. What to do? Put in the two centerpunch marks. Support one on the tailstock center, and drill the other end from the headstock. Note that you may need three hands, and also that if the workpiece is a smooth round bar, your hands will likely be all that's holding it from rotating with the drill, so PROCEED WITH CARE!! Maybe you think it won't work? It will - I've done it myself about half a dozen times. Obviously, start the drilling operation with a good sharp center drill, and progress up to the final drill size in small steps.

And here's a further variation on this one too: fix a "center" to your milling machine or drill press table exactly under the spindle centerline. Locate one end of the job on one of its centerpunched hole locations on this center, and drill into the other end. Swap ends half way through. The two holes will be lookin' right at each other. (One application: drilling a through-bolt hole in the buttstock of a two-piece gunstock.)

————=∇=————

PORTABLE POWERED BORING BARS FOR BORING LARGE CYLINDERS

A small machine shop I know of rebores a lot of air compressor rotor casings (or some similar part about the size of a 45 gal. drum or smaller) for most of the mines and pulp mills hereabouts. The shop does not have a lathe big enough to swing the workpieces involved, so the whole operation is carried out with the job laying buck naked on the shop floor like a dead pig, using some very simple welded up fixtures, and the cutter is driven by maybe a 1 HP electric motor hung on a banjo arm, yet the job is well done, and at a price better than that at which anyone else offers to do it.

How they do it is an interesting piece of know-how for hsm's - a smaller scale application may come along for you sometime.

They use what is generally known as a star boring bar. The drawing below shows the salient points.

Item B is a spider at each end of the job, and can be just a hefty chunk of plate plastered across the end of the job, or can be 3 legged, or whatever. These Spiders carry the bearings, which may be bushings, if for one time use; or ball races, if to be used over and over, as at the shop where I saw this done. The bearings carry a fairly skookum shaft, Item C, along which D travels without shake. D carries the toolbit, and a leadscrew nut for leadscrew E, the other end of which rides in a banjo plate at the far end of C. At every rotation of C, D, and E, the starwheel on E hits the projecting pin on B, rotating E through part of one turn, thereby advancing D a little way along C.

Setting up the Spiders so that the shaft C is concentric with the original bore of the job will obviously call for some careful work with indicator, wrenches, hammer, etc., but once you have a zero/zero reading at both ends, you can set the cutter to take the first cut, and flip the switch.

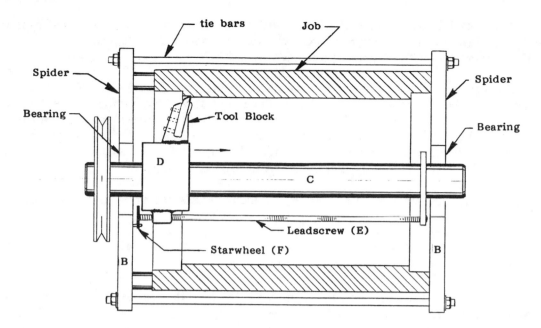

The cylinder of many a big steam engine has been rebored quickly and well with such a setup, the owners only too happy to pay a canny machinist handsomely, to avoid the king's ransom it'd cost to dismantle half the engine, and remove the cylinder to a distant machine shop for no better treatment on a horizontal boring mill.

A VISE ACCESSORY FOR HOLDING FLAT WORK
Adapted, with permission, from the
September 20, 1928 "WorkShop Topics" column
of *Model Engineer Magazine*

Anyone who does much fitting at the bench vise knows the difficulty of properly holding and supporting relatively long, thin, flat items for filing or polishing.

To grip a thin object by its edges in the bench vise poses obvious problems. The serrated edges of the vise jaws will mar the work. The vise can be fitted with a pair of relatively hard metal jaws with sharp, smooth edges; however, slop in the vise may thwart the intent. Further, even if one had suitable gripping edges for the vise, if the job is long and thin, the unsupported ends may bend downwards. One fix is to sweat the job to a flat (and thicker) plate; the snag is that you must completely finish all work on one side of the job before unsweating and reversing it to work on the other side.

The device shown here is useful when working on such parts, as it overcomes most of these problems. Thin flat work can be readily gripped tightly by its edges, with the top face clear for filing etc, and repositioning of the part takes only a moment or two. Jaw protrusion can be quickly adjusted to suit the thickness of the work at hand.

A similar item was available as a ready-made item many years ago. Altho it is called, among other names, an "instrument-maker's vise", it is actually not a vise as such, but rather an accessory for use with a bench vise. It is an excellent idea and a practical tool, but did not catch on, and was therefore dropped as a commercial proposition.

The commercial version was made with a cast iron Table, while the Jaws were forgings. In making a one-off version, one would employ somewhat different methods, which are reflected in the drawings given here.

As noted above, the workpiece will typically be some narrow and thin strip of steel; pressure is applied by the jaws of the bench vise acting on the Loose Jaw leg on one side, and on the back of the lug carrying the Fixed Jaw on the other.

The Pin in the horizontal Link between the lower end of the Jaws is inserted in whichever hole will put the Loose Jaw as near upright as possible for the width of a given workpiece. As drawn here, the maximum width of grip is just under 1-3/8".

In the commercial version of this tool, there was a flat spring (see dotted line, Fig. 1) attached with a countersunk flathead screw to the inside of the Fixed Jaw. This spring forces the Loose Jaw open when the bench vise is slacked. Although this is a slight convenience, the Spring can be omitted without much effect on the efficiency of the tool. If one is desired, it can be made from an old alarm clock spring, or whatever - it might be say 25 thou thick, or less, 1/2" wide where it attaches to the Fixed Jaw and tapering down to 3/8" at the tip.

Fig. 1 is a General Arrangement drawing of sorts; pertinent details not given there appear on the other drawings.

Fig. 2 is a Plan View of the Table, at about 1/3rd full size. Fig. 3 shows two views of the Table and Lug, with dimensions for placement of the Lug.

Fig. 4 shows a detail of the Pin and the slot in the lower end of the Loose Jaw, the latter being marked "a", while the horizontal leg of the Fixed Jaw is marked "b", as they also are in Fig. 1.

Make the Table from 5/16 or 3/8" HR plate. Mill to overall dimensions, mark out the slot in the Table, and match drill the holes from the Table into the Lug, which is made from 1/2 x 1" CRS.

The edges of the Table can later be given a neat 1/16" 45° chamfer with a milling cutter such as that referred to in the Footnote at the bottom of Page 169 of **TMBR**, or by other means, and when all else is done, the Lug can be screwed to the Table. If a permanent assembly is desired, the joint can also be silver soldered; if you do this, make a nice job of it, leaving only a fine fillet of solder visible at the root of the joint.

The Jaws can be made of annealed 1/2" square drill rod, or 1/2" sq. hot rolled tool steel; they should be machined to spec, nicely polished, and hardened & tempered to just below the working portion. Alternatively, make from 1/2" sq. CRS, or keystock, and caseharden the working portion.

[Footnote to the above, from the guy who brought you that incomparable piece of technical literature known as **The Bullseye Mixture**:

When the original author said ".. and caseharden the working portion.", he was no doubt thinking in terms of the use of Kasenit. I'd be inclined to caseharden both jaws in their entirety, as well as the Lug, which, with the Loose Jaw, is subject to the bite of the serrated bench vise jaws (if they be not covered with soft jaws of some sort). I would then wrap the gripping part of the Fixed Jaw in a wet rag to maintain the hardness of the top end of same while brazing in the Link piece (see below). GBL]

The Link is made from a 2-1/2" length of 1/8 x 1/2" CRS - this is not called out on the drawing. Use a slitting saw to make a 1/8" wide slot in the lower end of the Fixed Jaw; drill* and pin the Link and the Fixed Jaw together, as shown, and then braze the joint solid. File excess brazing material back to the original surfaces thereafter for a neat appearance.

*Obviously, the corresponding hole in the bottom of the Fixed Jaw will be drilled before you caseharden same; otherwise the hard case will have to be annealed before you try to drill the holes. GBL

The Loose Jaw is straightforward. It should operate freely without fouling or binding in the slot in the Table, and must be so made that, when upright, it stands proud of the Table the same amount as the Fixed Jaw.

Fig. 1

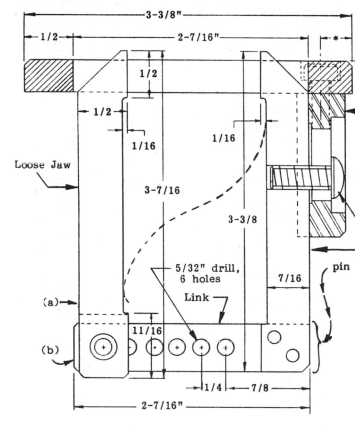

Loose Jaw

3-3/8"
1/2
2-7/16"
*
1/2
5/16"
1/2
1/16
1/16
1/2
3-7/16
3-3/8
(a)
5/32" drill, 6 holes
Link
(b)
11/16
1/4 7/8
2-7/16"
7/16

Fig. 2

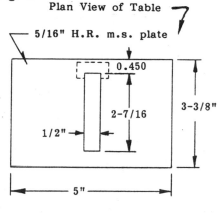

Plan View of Table

5/16" H.R. m.s. plate

0.450
2-7/16
1/2"
3-3/8"
5"

Lug

1/4-28 button head cap screw

Fixed Jaw

pin & braze

*In the sectioned side elevation drawing at left above, note that the socket head cap screw (one of two; shown by dotted lines) which hold the Lug to the underside of the Table IS NOT CORRECTLY LOCATED. It is shown located on the centerline of the 1/2" thickness of the Lug. In fact, it should be located per dimensions given in the drawing at lower left of this page.

Pin Detail:

1/4-28 x 3/16
1/2"Ø

Fig. 4

1/2

Loctite

(a)
(b)
5/32" drill rod, 7/8"OAL

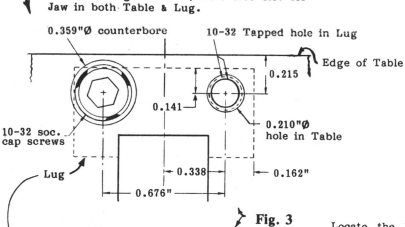

Scrap detail showing placement of screws to attach Lug to Table; note also slot for Jaw in both Table & Lug.

0.359"Ø counterbore   10-32 Tapped hole in Lug

Edge of Table

0.215

10-32 soc. cap screws

0.141

0.210"Ø hole in Table

Lug

0.338

0.162"

0.676"

Fig. 3

Slot for adjustment of jaw protrusion

3/4"
1-1/2"

1"

Slot, 0.475 x 0.140 x 1", for head of 1/4-28 button head cap screw.

Locate the holes in the Table by means of accurate layout work, then drill & counterbore for the two 10-32 shcs's. Locate the corresponding holes in the Lug "by the numbers", when these are then drilled & tapped, the two parts should go together like a hand in a glove. After checking that this is so, mill the 1/2 x 1/8 channel in the "Jaw" side of the Lug. The Fixed Jaw slides vertically in this channel. Make the channel say 3 thou over the nominal 1/2" width. Relieve the floor of this channel about 20 thou in the area of the 1/4"-clearing through-slot.

Mark out the 1/2" slot in the Table; drill holes in the corners of the slot to minimize later file work. Mill the slot, and file the ends of the slot square with a safe edged file (pillar file).

As for the holes in the knuckle at the bottom of the Loose Jaw, the original writer called for the use of a taper pin reamer in the hole, and a tapered Pin to match. I would use drilled holes in both parts, and a 5/32" straight pin; in making the latter, I would want to end up with an easy but not sloppy fit in the drilled holes - somewhere between 5/32 and #21 drill should be about right. Drill the corresponding 5/32" hole at the bottom of the Loose Jaw, then open out to #3 on one side, and tap 1/4-28. (All this will of course be done before casehardening; it might be a good idea to plug the threaded hole with a "service screw" prior to casehardening, this screw to be removed and garbaged when the job comes out of the quench tank after casehardening.)

The Pin is straightforward, and is most easily made as drawn. One could make the Pin and Knob in one piece, but there are disadvantages to this: The application is such that, for long years of use, something better than a mild steel pin is called for. We could make the Knob and Pin in one piece from say 1/2" drill rod, but drill rod does not knurl easily - it's tough on knurling wheels. However, this might not be a problem in this case, so long as the portion to be knurled is narrow, hence lotsa pressure from the knurls on a small area. But don't dawdle about the matter: as noted at page 61 in **TMBR**, always get a knurl on in as few turns of the work as possible; this would be doubly true in knurling drill rod.

Footnote: A modification of the foregoing tool idea appeared in *M.E.* for June 7/74. There, the author put legs about 4" long at the 4 corners of the table, and put a bottom plate under them. He then added a vise screw etc., so the gadget was a vise in its own right. He thus had a useful item for holding thin, flat work in the drill press, or like that.

BEWARE THE MAN WITH MORE THAN ONE HACKSAW

You've heard the expression "Beware the man with one rifle - he probably knows how to use it."

Well, in hacksaws, the opposite is true. I used to wonder why Bill Fenton had about half a dozen hand hacksaws in his shop. It was probably while switching blades in my own hacksaw one day that the answer came to me: each of Bill's frames has a different type of blade - fine tooth, coarse tooth, brass only, steel only, etc. Bill wastes no time changing blades. A luxury? No, a smart convenience, and not expensive. You can probably pick up hacksaw frames very cheaply in second hand stores, pawn shops, at garage sales, and so on - everybody else thinks that if they have one they are well equipped, so why buy another? Well, now you know.

"A REAL MAN'S HACKSAW"

And if you want "a real man's hacksaw", deep in the frame, and rigid, you can make one up along the lines shown below. I saw this in the shop of a canny old machinist by the name of Tom Coey. Tom has a couple of portable 2-cylinder power units he made using cut-in-half Ford Model T engines. They run as smooth as can be, and are a neat and workmanlike job in every respect. Tom takes these engines to antique machinery shows, and they attract much interest.

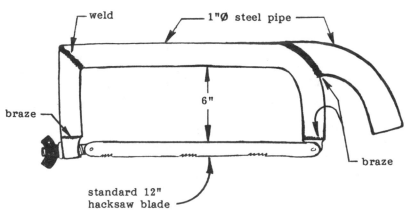

100

SHOP MADE ANGLE PLATES

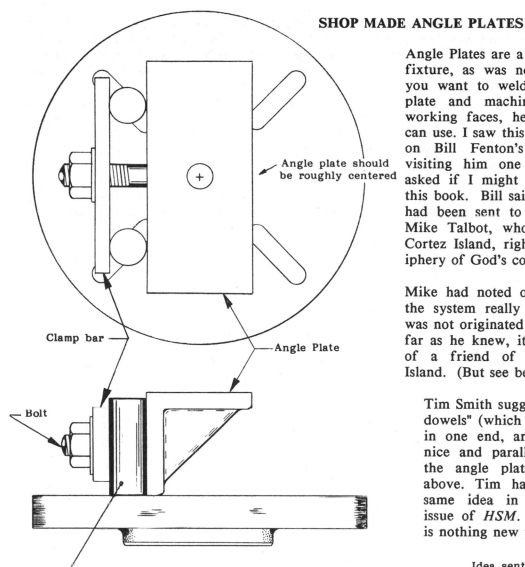

Angle plate should be roughly centered

Clamp bar

Angle Plate

Bolt

2 posts, ends faced off square to axis, O.D. uniform full length - make from CRS, or Drill Rod. Secure to faceplate via bolts/washers from back side of Faceplate.

Angle Plates are a very useful type of fixture, as was noted in **TMBR**. If you want to weld one up from HR plate and machine it on the two working faces, here's one setup you can use. I saw this on a sheet of paper on Bill Fenton's workbench while visiting him one day recently, and asked if I might work the idea into this book. Bill said yes, noting that it had been sent to him by his friend Mike Talbot, who hangs his hat on Cortez Island, right here on the periphery of God's country.

Mike had noted on the drawing that the system really works, and that it was not originated by him, but that as far as he knew, it was the brainchild of a friend of his on Vancouver Island. (But see below!)

Tim Smith suggests the use of "pull dowels" (which have a tapped hole in one end, and which are made nice and parallel by grinding) in the angle plate machining setup above. Tim had noted much the same idea in the May/June 87 issue of *HSM*. Yea, verily, there is nothing new under the sun.

Idea sent to Bill Fenton by
J.M. Talbot, Whaletown, B.C.

SOME HANDY TOOLS
from a letter from
Jobie Spencer of Urbana, Illinois

Eclipse Tool (an English outfit) makes a small hacksaw with which they supply 3 grades of blades, 32 teeth per inch, 44tpi, and 60tpi. I have never seen anything better for making smooth cuts in thin wall tubing, heat treated socket cap crews and the like. The handle is well shaped for good control, and the blades are thin - about 20 thou. The Part Number on it is "#45". Some of the woodworkers' supply outfits carry it, and the price is right down there where you can get at it without putting your wallet outta joint, too.

(NOTE: Coles' and Mittermeier's catalogs both show such a hacksaw, and blades for same. You might want to buy the saw, or just the blades, and thereafter make a frame to suit - see below. GBL)

I (Jobie) have also made a number of small hacksaws per drawing below. It follows a design which I believe was originated by a friend of mine, Ernie Ditzler, who to me embodied all that a machinist should be.

He had been a coal miner, and a tool & die maker, and he had made some of the tooling and drilling jigs for Allison engines during WWII. I worked with him for many years, and learned a great deal from him - in fact his was the best training I have ever had. Not formally educated beyond the 8th grade, "Mr. D", as he was often called, was a very smart individual, as witness the following:

When he applied for the job of Instrument Maker at the University of Illinois (where I eventually met him), he had to write an exam which included several math problems. Only after he'd finished them all did he realize that he was allowed to use paper and pencil for his calculations - he had done them all in his head!

Anyway, back to the little hacksaw: it will use blades from the Eclipse saw mentioned above. I have made several of these and given some to friends. The drawing below is self explanatory.

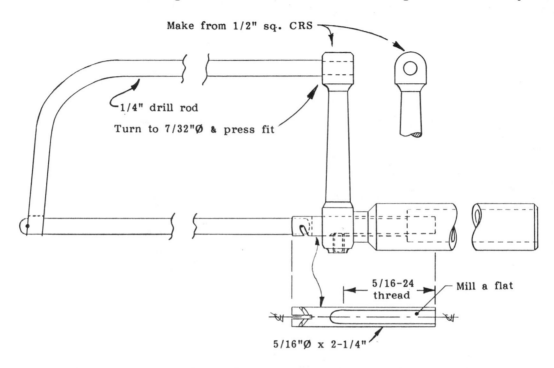

Make from 1/2" sq. CRS

1/4" drill rod

Turn to 7/32"Ø & press fit

5/16-24 thread

Mill a flat

5/16"Ø x 2-1/4"

(Insert from GBL: right here is a handy place to insert a drawing of a small saw frame a fella could make for himself for doing fine work. Carve the back end from a piece of 1/4" HR plate; forge or weld on a tang about like on a file, to go into a (bought) file handle. The bent part of the frame could be made from 1/8 x 1/4" CRS or slightly heavier material, bent cold or hotted up and forged to shape. Use with jeweler's saw blades.)

NOTE: commercially made saws of this type use an even simpler form of clamp than shown here - they omit the ridges at heel and toe of the clamp shoe.

Footnote to the 2nd printing: Ted Lewis of San Bernardino,CA., who knows much about using such saws, recommends putting the handle lower on the frame - in fact directly in line with the blade axis - to reduce the tendency for blade breakage.

Jobie's letter then went on to detail several other tools which are worth noting here. Of them he says:

....I'm sure any machinist worth his salt can and has thought of the ideas here, but these are things I use all the time in large or small work. For an occasional hole or two it might not be worth the time to make some of them, but for continuous work and time saving they can't be beat, especially if you are using a larger lathe for small work.

Note that a number of these tools employ dowel pins in their construction. Dowel pins in assorted sizes (long) are very handy, and are the straightest and most true to size article I have found (better than drill rod or drill blanks). Also inexpensive.

(I have heard that the needles from needle roller bearings are cheaper than dowel pins, and equally good for quality. I don't know if this is true or not, nor if needle rollers would be suitable everywhere a dowel pin would be first choice otherwise, but it is something to keep in mind. GBL)

Tool Extension Shank

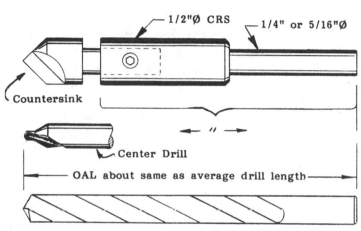

1/2"Ø CRS   1/4" or 5/16"Ø

Countersink

Center Drill

OAL about same as average drill length

It doesn't make much sense to raise the drill press table to center drill a hole, lower it to drill the hole, then raise it again to deburr or countersink. Also useful for hand-holding suitable cutting tools for hand deburring, etc. I've made lots of these, and had an apprentice make a bunch one time. Make in whatever sizes are useful.

I thought of it, but others have too - I saw something similar in Frank McLean's column, one time. Not for hi-precision work in the lathe, but fine for the class of work usually done on a drill press.

Tail Stock Die Holder

There are many versions of these, but I've never read an adequate description of how to use one. Make in sizes to suit your dies. Those for larger thread sizes can be held directly in the tailstock chuck; the ones for the smaller sizes of threading dies can be slipped onto a piece of 1/4" drill rod or a HSS drill blank (or a long dowel pin, which is just about perfect) held in chuck. This allows the die holder to be held and fed by hand. When the thread is deep enough, just release your grip on the holder and it will spin on the 1/4" rod until the lathe is shut off. Saves breakage of work or tap, and gives precise control. Holes in the side allow you to apply cutting oil, and watch the progress of the work.

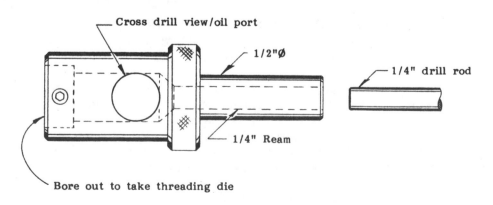

Cross drill view/oil port

1/2"Ø

1/4" drill rod

1/4" Ream

Bore out to take threading die

Extension for a Small Drill Chuck

Another of Mr. D.'s good ideas: this gadget provides small hole drilling capability on a larger lathe. It can be held directly in the tailstock chuck, but the best way is with a slip mandrel of 1/4" dowel or drill rod. It can be manipulated with fingers for sensitive hole drilling or tapping; held lightly and released at the proper time, it will spin until the lathe is turned off or reversed. Several sizes are useful.

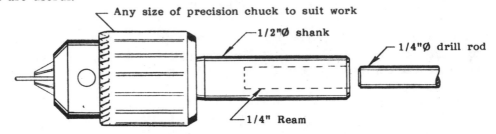

— Any size of precision chuck to suit work
— 1/2"Ø shank
— 1/4"Ø drill rod
— 1/4" Ream

Good Countersinks: The best I've found are made by Ford and I prefer them to anything else. They make single flute countersinks down to at least 1/8" dia. and I have several up to 1". The 5/8" size is very useful, as it has a 1/4" shank.

A Center Finder

Make from dowel pins. Several sizes will be useful - sizes I use most are 1/4", 3/8", and 1/2". Use in machine spindles for locating cross lines. For greatest accuracy run spindle at low speed and use a magnifier. Quickly locates center to within 0.005.

... and some more tricks for the Oddleg Artist: For layout work I use steel rules, squares, scribes - I never did own a height gage, though I had access to one if I needed it. I have built a number of complex machines and most of the prototypes for Rhino Robots, Inc.

When cutting out parts on a mill, I square the pieces up first, then finish layout lines, etc. using a square set with vernier caliper. Then cut to line by eye, first cutting down to say 0.032" (1/32") which is fairly easy to estimate. The last cut is made by estimating the remainder, and I generally would be within 0.005" or less without any measurement at all, provided the original layout lines were accurate to start with. This approach, where acceptable, makes work go fast.

A Shop-made Scriber

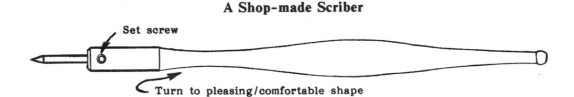

Set screw

Turn to pleasing/comfortable shape

Use a broken HSS tap for a scriber tip. It will cut through mill scale pretty well and not get dull too fast.

OIL SQUIRTERS FROM SHAMPOO BOTTLES ETC.
Thanks to Mr. R. Kitchen for permission to show here an idea he reported in the Aug 21, 1970 issue of *M.E.*

You can make handy little shop squirter bottles from the variety of plastic hair shampoo, aftershave lotion bottles and so on which pass through the average household in the course of a year. Turn up a shouldered brass plug a good snug push fit in the neck of the bottle, and machine the outside/nozzle profile to taste. The plug is drilled full length and the business end can be threaded for a screw cap, if desired, making the bottle spill proof, maybe even leakproof. Bore the bottle's original cap to clear the brass nozzle, as drawing, screw it on, and you're done.

You can also solder a piece of brass tubing of any length desired into the brass plug; this economizes on material and saves having to drill a longish hole in the spout. This type is particularly handy when boring deep holes, and on other jobs where one wishes to deliver the fluid to an otherwise inaccessible spot.

When you first take one of these bottles out for a test drive, be careful until you get the feel of it - too hefty a squeeze can send a jet of thin oil clear across the shop. Note also that some cutting fluids (e.g. "Rapid Tap") have adverse effects on some plastics.

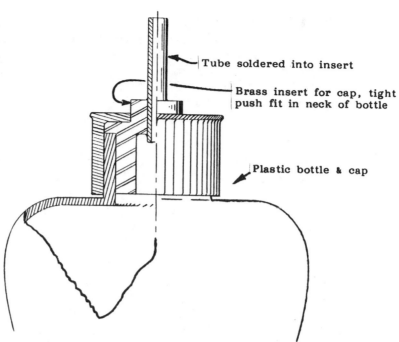

Tube soldered into insert

Brass insert for cap, tight push fit in neck of bottle

Plastic bottle & cap

RADIUSING THE END OF A PART

Suppose you have to make a part as at A below. How to machine the radius on the end? It is not a job that lends itself well to being done on a rotary table - the cut is so broad that you will not likely get a nice, chatter-free cut and a good finish off the tool.

Best way to handle this is to blue the work, make your layout with centerpunch, scriber, and dividers, and then drill/ream/bore the hole. Then hang the workpiece on a piece of rod laid over the top of your milling vise, as at B. Grip the job solidly in the vise, and mill a flat right down to the scribed line. Set your quill depth stop, then up quill, turn off machine, and re-orient the part slightly. Down quill, and mill another flat. Repeat this until you've gone all the way around the curve, and then start milling off the points between adjacent flats, until the flats are quite short, as at C.

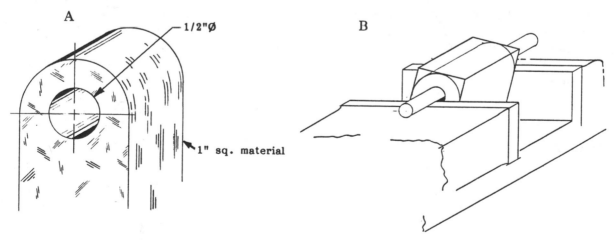

A

1/2"∅

1" sq. material

B

NOTE: turn the machine off between moves.
Let me repeat that: *Turn the machine off between moves.*

Obviously, while doing this, you will need to remove the workpiece frequently to wipe everything clear of chips, lest a chip get under the rod and mess you up.

Less obvious at first may be that the rod on which the workpiece is hung needn't be a close fit in the hole in the job - a 1/4" rod in a 1/2" hole is fine.

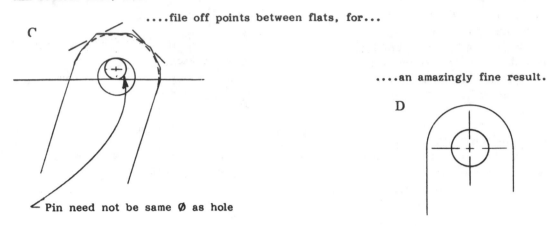

Mill several flats, and...

....file off points between flats, for...

C

....an amazingly fine result.

D

Pin need not be same Ø as hole

Now, when you've milled about as many flats as you feel like, unship the job from the vise and go to work on it with a file, smoothing off the profile. The width of the workpiece lets you keep the file from rocking from side to side. The pressure of the file on the work is highest when it is cutting on the points between flats, hence the points are cut down very quickly, while little is taken off when the file is parallel to any given flat. You'll find you can file the job to the desired profile much more easily than you would believe until you try it. Your fingers will tell you more than your eyes about your progress, and you will say to yourself, "By gummy, that guy Lautard knows what he's talking' about!"

FILING BUTTONS

The foregoing brings up another useful aid - the filing button. This is a simple idea, but it works well. Say you want to file a radius at each of the four corners of a metal rectangle. The job could be anything, but let's say it's a "builder's plate" you're making for a model you've built. Having engraved it on your bench top engraving machine, you want to radius the corners. Say the radius is to be 1/4".

Face the end of a piece of 1/2" drill rod, and part off a slab about 1/8" thick. Make a second one, just like the first. Deburr and harden both pieces. Clamp to the job, one on either side, positioning against a couple of surfaces at right angles (e.g. the face of your 3-jaw chuck and the side of one of the jaws) so the buttons are flush with adjacent edges of the plate. Now file the exposed metal down to the buttons. The file will skate on the buttons, because they are harder than the file. When all four corners are done, a couple of strokes with fine abrasive paper and you have a first class job.

The same idea can be applied in many situations. Don't overlook the possibility of securing the buttons in place via a screw, one button being clearance drilled, the other tapped. Also, one might stick the buttons on with adhesive, if the workpiece doesn't/can't have a hole through which the screw can pass. Heat or freeze to remove.

Filing buttons need not be round - they can be hex, or any other profile you want to duplicate nicely by filing. Buttons can be filed as a pair to any desired profile, in say 1/8" sheet steel, and then casehardened. If not hardened, greater care will be required in use, to avoid cutting into the button.

And just because the file skates on a hardened button doesn't mean you can bear down on the file all the harder to get the job done faster - that hard steel is not doing the teeth of the file any favors, so go easy!!

Not all jobs will require, or allow, paired buttons - sometimes one, on one side of the job, will be enough, although greater care may be required to keep the file square.

One day in my mail I received, from Tim Smith, a copy of an article about the Paige Compositor - a typesetting device developed in the 1880's - and its arch rival, the Linotype machine, which was originated by one Ottmar Mergenthaler, and as I understand the article, eventually came to be manufactured by - The Mergenthaler Linotype Company.

(Ultimately, the Paige Compositor was overshadowed by the Linotype equipment because Paige sought perfection rather than a workable design that could be made and sold. Had Paige's financial backers wrested control from the inventor and put the machine into production, they would have had a winner on their hands, because it was several times faster than the Linotype machine. But all that is another story - if you are interested, get a copy of pages 55 through 60 of the Summer 1987 issue of the magazine INVENTION & TECHNOLOGY.)

Anyway, along comes this interesting article, and the next piece of mail was a letter from *American Machinist* Magazine granting me permission to reprint the article which follows. So I coughed it up on the screen of my computer to note permission had been granted, and..... notice the employer of the author.

A FIXTURE FOR ROUNDING THE ENDS OF SMALL PARTS
by Henry W. Boehly
Machine Designer, Mergenthaler Linotype Company
(reprinted, with permission, from
American Machinist Magazine, March 5, 1931)

When building special machines on contract, analyzing the job to find the best method of machining the parts is very important, especially if the cost is to be kept out of the red. We had to make several dozen only of parts like the one shown at right. Since the quantity was small, and the dimension A was variable, making press tools for the job was out of the question.

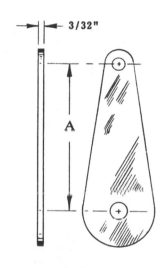

A strip of stock of the same width as the length of parts was cut from a sheet in the usual manner. The blanks were cut from the strip in a power shear by feeding the strip at the required angle and alternately flopping it over for each piece. The holes were pierced in a set of press tools that were always kept in working order, for piercing holes in similar work was almost a daily occurrence. It was necessary only to make a new nest for the piece.

Face milling cutter in lathe spindle nose

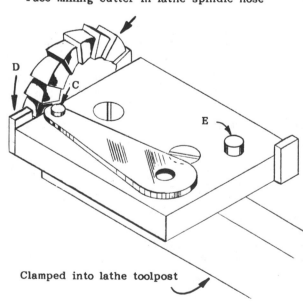

Clamped into lathe toolpost

After reaming the holes, the ends were rounded in the lathe, using the fixture shown at left. This fixture consists of a shank to fit the toolpost, and on the top a removable plate containing two pins. The small end of the piece was slipped over the pin C and fed up to the cutter, which was held on an arbor in the lathe spindle. By slowly swinging the piece by hand until it contacted the stop D, one-half of the end was rounded. The piece was then turned over on the pin and the other half was rounded in the same manner. After the small ends of all the pieces had been rounded, the plate containing the pins was removed and replaced in the reverse position, bringing the pin E nearest the cutter. The large end of the piece was put over this pin, and its end was rounded in the same manner. A stop on the lathe bed served to locate the carriage for the correct depth of cut.

A LITTLE PIECE OF NEW TECHNOLOGY

I had the pleasure of touring a research machine shop recently - interesting to see how the other half lives! One chap was carving out something or other from a piece of 1-1/2" dia. stainless. The first of two was done and sitting on the bench beside his mill, and next to it was a piece of what looked like red-brown steel wool.

I asked him what it was. He said it was "ScotchBrite", an abrasive plastic scratch pad. Looks sort of like a furnace filter. The fella said to try it on the workpiece, so I did, and be darned if it didn't produce the nicest "brushed" finish you could ever want!

Turns out this stuff is a 3M product*, the coarser grades of which can be bought in most any supermarket, where it's sold for scouring pots and pans, cleaning floor tiles, etc. The finer grades (and, I would expect, the coarser grades too) are sold by tool/industrial supply outfits. The grit size seems to go by color - the green pads are coarser, and the red-brown stuff is about perfect for general finishing; grey is extra fine.

*3M is not the only outfit that makes abrasive materials of this type; other companies make similar products.

———————▽———————

A LATHE TRACING ATTACHMENT

Readers will recall the section in **TMBR** on turning balls and other profiles without specialized attachments. Tim Smith sent me details of a simply made attachment for your lathe which uses a different approach but which is equally versatile. In the October 1951 issue of *Popular Science*, at page 117 and 168, you will find full working drawings, but I will tell you the principle here.

The profile to be produced on the work is drawn full size, either on paper, which is then used directly (as described below), or transferred to a piece of metal which is then filed to shape, for a more permanent template. The template - paper or metal - can be attached to the lathe, with its axis lined up with the lathe's axis. In the original article, the template was shown attached to the tailstock barrel. This is a convenient place, but there is no reason it could not be clamped to the bed of the lathe or elsewhere, if more convenient.

A pointer is then clamped solidly to some part of the lathe compound rest. This pointer will want to have means of universal adjustment. Either clamps as found on a surface gage, or other, lo even unto the simplicity of a heavy, bent wire. Whatever is used, the working end should come to a rather sharp point.

To produce the desired profile on a piece of stock chucked and already turned to the intended max. OD, juggle the cutting tool and the indicating tip of the pointer into appropriate positions on the work and the template respectively. Then, carefully watching the proximity of the pointer to the edge of the template, and with one eye on the workpiece and one on the toolbit, work the carriage, topslide, and compound feedscrews to maneuver the pointer along the template. Finally, smooth out the turned profile of the workpiece with a file, and/or file-backed abrasive paper.

Bob Haralson tells me that in one shop in which he worked shortly after WWII, they frequently had to cut cam-like profiles along the edge of a piece of steel plate. The cam profile would be laid out on the material, which would then be clamped to the table of the vertical mill. The machinist would juggle the two table feedscrews to cut the desired profile to the layout line using the side of an end mill. Bob said this was done routinely and with excellent results. Mind you, this was mostly for sawmill machinery, not wrist watch parts, but it still shows you what clever people machinists are: who else could operate two cranks with two hands, each hand operating its crank in either direction, totally independent of the other?

This came up one day when Bob and I were working on something together; I commented on having noticed that I had acquired the ability to do this, and Bob's reply was "...everybody who works with machine tools very quickly learns that trick." Shucks - I thought it was just my native inteligents.

Back to the matter of producing a ball profile in the lathe: In one of his letters, Tim Smith told me about watching a tool & die maker at work use a washer with a trued up hole to locate the high spots on a ball he was free hand turning to shape. He didn't say if the hole was the same size as the ball, or smaller - either way would work, I think. **Also:** If the washer was hardened, and the hole made sharp edged, and about 2/3rds the Ø of the ball desired, it could be held, *carefully*, in a gloved hand, and used as a scrape cutting tool to produce an almost perfect ball.

A SIMPLE STAMPING FIXTURE

We sometimes need to stamp a row of numbers around a graduated dial made in the shop. If you want the impressions to be nicely lined up, it is, for most of us, imperative to use some type of fixture to hold the stamps, and another to hold the work. Illustrated below is a stamp (number punch) holder that can be rigged for use from the lathe toolpost (for a job still in the chuck) or from the milling machine spindle nose, in the case of a job which has been moved to a dividing head set up on the mill. Although the holder must not move, it need not be extremely rigid in its mounting, because the stamp, and not the fixture, is what gets hit.

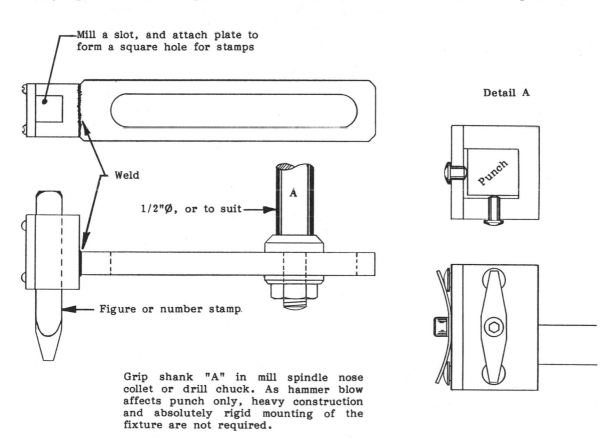

—Mill a slot, and attach plate to form a square hole for stamps

Detail A

Weld

1/2"Ø, or to suit—

A

Figure or number stamp.

Grip shank "A" in mill spindle nose collet or drill chuck. As hammer blow affects punch only, heavy construction and absolutely rigid mounting of the fixture are not required.

A further refinement is shown at Detail A. Here, 2 pairs of button-headed pins are spring loaded with leaf springs. The latter can be made from a piece of steel strapping of the sort found in garbage bins behind many commercial establishments. The pins are shown in the Plan view of Detail A, while the leaf spring appears in the Elevation view. Note that the pins are shown sprung outward slightly by the punch, which is shown in place in the Holder only in the Plan view. This arrangement works well - I once made such a device which would fit in a 3/4"Ø reamed hole in the upper arm of a G.H. Thomas-designed tapping & staking tool. (If interested in making such a tool, which will be found to be a valuable and versatile tool for this and other types of work, and which has the ability to serve also as a small, sensitive drill press, write to Neil Hemmingway in England for details - see Appendix for his address.)

Two facts to keep in mind:
1) Not all characters are properly centered on their stamp's shank, even on good quality hand stamps. They may not be far out, but unless this is corrected by surface grinding the offenders, even the use of a fixture will not give you the perfection of alignment you may seek.

2) Different characters require blows of different force to make uniformly deep impressions - for example, a "1" (one) requires less force than an "8".

Also, as was noted in **TMBR**, for a nicer final appearance, pass a file over the work after stamping, to dress down the metal thrown up around the character impressions.

SIMPLE SHEET METAL BENDING DEVICES

I haven't done a lot of sheet metal work, but I do like working with the stuff when I have reason to. Nice smooth edges, filed free of burrs before bending, and nice clean, uniform bends, careful work with hammer, blocks, etc. to adjust a bend if need be, and all like that there - very satisfying. For the little sheet metal work I have done, I have simply clamped the work between a couple of pieces of CRS in the vise, and possibly clamped the middle and/or outboard ends of same with a good size toolmaker's clamp, and then used a block of hardwood and a soft faced hammer to effect the bends.

You can make one or both of the items shown below for sheet metal bending. These are very simple, and need little or no comment. If you want to bend up a series of like-sized lips, sandwich a strip of material the same thickness as the material to be bent in the jig as indicated by the hatched piece, and set it as a stop, so material goes in to the same depth for each bend.

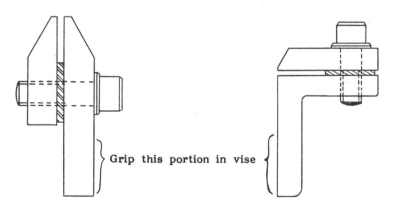

Grip this portion in vise

Don't forget to allow for material consumed in the bend. See "Bending allowances, sheet metal" in *Machinery's H'book* for details on how to figure this out.

Here's a little story on the subject of sheet metal work:

My friend Bob Haralson said that during his training as a machinist at Aberdeen Proving Ground in the mid '40's, there was a course in sheet metal work which he very much wanted to take, but was unable to. It was taught by a fella who was an absolute genius with sheet metal - witness the following: One day an officer came into the Base Metal Shop and handed this chap the clip (magazine) from a M22 Springfield .22 rifle. Could he make him another like it? "By the end of the day, that fella had made a new clip, and when you laid the two side by side, the only way to tell them apart was that one had the Springfield Armory markings stamped on it. And this he did while supervising the work of maybe a dozen other men."

It might bear mention that a clip from a M22 Springfield rifle has a depression - said depression not being just a simple rectangle, either - stamped in both sides, leaving a couple of ribs running up the rear of the sides; there is a curved slot in the right side, and a grooved-head button comes thru this slot, presumably attached to the follower; and the floorplate is not just plain flat, either: it has a series of wrinkles in it. It would not have been a simple item to duplicate on a one-off basis.

Late addition to the above: Just happened to spot, in a custom automotive parts store, a book entitled *Metal Fabricating Techniques for the Custom Car Builder*, by Ron Fournier. This book is about making whole new body parts, shapes, etc. from scratch, in aluminum, steel, and other sheet metal. If you have an interest in knowing how to form larger pieces of sheet metal into complex curves, do see this book - it deals with welding, clamping, forming tools, and much more, and it is good. You could probably get it at your local library.

Smith goes into Orbit

Tim Smith wrote me one day saying he'd tried his orbital sander on a piece of stainless steel sheet he was working on. He said it produced, with almost no effort, the most beautiful finish you could ever desire.

A TOOL TO AID NICELY FINISHED LATHE-CUT THREADS

I suspect a lot of hsm's avoid screwcutting like the bubonic plague. If one lacks a quick change gear box, setting up a train of gears to cut a particular thread is a time consuming process, and may be a factor in why some avoid screwcutting as much as possible. But I think a lot of guys who avoid it do so because they're scared to try. Screwcutting is not difficult. Cut a few throw-away pieces, and you'll acquire the initial experience that may be holding you back from doing it routinely.

I'm not going to launch into a dissertation on how to screwcut - that is amply covered in any decent book on basic machine shop practice. I will however tell you a few tricks, some of which you may not find in that sort of book. (Most of what follows is not aimed at the man racing the clock to turn out coarse threaded bridge bolts, but rather the guy who wants to cut a real nice "toolroom grade" thread on something he's makin' for himself.)

1) Have you ever wondered why the lathe compound is so often shown set over to 29° (or 29.5°) instead of 30° in many text books showing how to screwcut? I puzzled over this for quite a while, and I think I have finally tumbled to the answer: the screwcutting toolbit is a form tool, presumably accurately ground to 60°. By setting the compound at 29-1/2°, we get the advantage of the down-the-flank infeed method, while at the same time we allow the form tool to give us the correct shape of the thread. That's all there is to it! If you don't see it on first reading, read the last two sentences over again a couple of times, and all of a sudden it'll come to you.

2) Every so often when cutting a thread, make a pass or two without advancing the tool - like as not it'll take a little more metal off. This works out any spring in the workpiece or the toolpost setup.

3) Every so often when cutting a thread, advance the cross slide by 0.001", and make a pass without advancing the compound slide. Then make the next pass with the cross slide feedscrew dial re-set to zero (i.e. 0.001" back from previous pass) and advance the compound slide enough to cut again.

4) Try taking a pass down the thread while putting just a little drag on the carriage handwheel. This has about the same effect as 3) above.

5) When starting to cut a thread (and after checking that you are in fact going to cut the pitch you want), take a cut of about 0.005" deep, followed by one of about 4, then 3, and so on, not forgetting 2) above. The final cuts should be with tool advances of 0.001" or less.

6)   If you are not getting a nice finish on a broad cut such as is involved in the use of a screwcutting tool,

     a) make sure your toolbit is sharp - hone it with a hand stone as discussed elsewhere herein, if need be, and

     b) if a) is in order, drop your spindle speed substantially, and slather on some thick cutting oil. Make your last passes by hand power if need be.

> (TIP: In this regard, a useful lathe accessory is a mandrel hand crank which can be plugged into the outboard end of the lathe spindle, and locked there by means of an expanding plug. Just don't leave it in there when you run the mandrel under power again!)

7)   There was a time when it was commonly thought that good threads could not be cut on a small lathe without the use of a spring type toolholder. I don't use such a toolholder, but I can see why it would aid the cutting of well finished threads: its design is such that if the cut is heavy enough to spring the tool, the tool will spring away from the work, reducing the cut, and easing the spring. This makes sense, and if I had trouble producing nice threads, I'd make myself such a holder. You may want one.

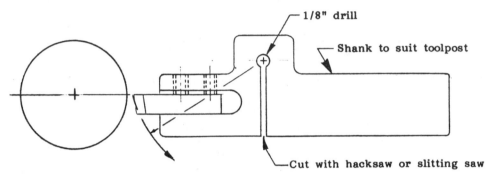

Tool pivots down and away from
work if cut is too heavy

> TIP: Make the cutting tool from a 1/4"Ø HSS drill blank (or 1/4"Ø drill rod). Ream hole for toolbit in end of spring toolholder, and fit dual 10-32 socket set screws with copper, brass, or lead pads to bear on the cutter shank.

> TIP: To reduce the springy-ness of this toolholder, stick a piece of wood, leather, plastic or whatever seems best into the bottom of the slot.

How much Infeed on the Topslide?

Here's something you may want to paste up near your lathe, and/or on or near your drawing board.

> Divide the number of threads per inch into the number 0.75 to get the infeed to apply on the topslide (or compound feedscrew when feeding in at 30°. If using the 29° set-up, use 0.742 instead of 0.75.

Tim Smith tells me the above works real slick for him. This is the sort of stuff that is easy to calculate and note down when doing working drawings, and thus save time in the shop.

A KNOCKIN' BLOCK

Friend of mine has a block of lead, a little bigger than an egg, which he uses to tap jobs into solid contact with the chuck face in the process of chucking them for machining. From its appearance, he uses it plenty, which is evidence of its utility.

Such a block is easy to make, if you have some sort of mold to cast it in, and a source of lead, a pot to melt it in, and a source of heat.. all of which is easy to get:

For a pot, get a bullet caster's lead melting pot; go to most any gun shop, or weld one up from a piece of pipe and a piece of plate. Used wheel weights, obtainable at a tire store, are a handy source of lead.

You can melt lead on a gas-fired kitchen stove, or a Coleman camp stove, or similar. (In the interests of domestic tranquility, I suggest you go with the camp stove.)

WARNING SPECIAL NOTE: Heat the ladle thoroughly before putting it into the molten lead.

Molten lead typically has a temperature of about **750°F**. **Be absolutely certain** there is no water or moisture in anything - additional lead, the ladle, whatever - that you put into the molten bath of lead. If there is moisture in anything added to the already melted lead, the water will flash to steam in an explosive manner, and **you are almost guaranteed to be spattered with molten lead**. Or to put that another way, you are almost certain to receive terrible facial burns. 'Nuf said.

Put about half a teaspoon of lard onto the molten lead to "flux" it; stir it, and skim off the wheel rim clips and all the dross that floats to the surface. Then carefully ladle the bright molten metal into your mold.

This mold can be a simple wooden affair, with an open top. Make the cavity say 1-1/2 to 1-3/4" square; it would not be a bad idea to give the cavity a bit of draft to ease extraction of the cooled lead block. Also, although the mold's joints do not need to be gas tight, you can line the cavity with brown wrapping paper to keep lead from flowing into the cracks between the pieces of wood that comprise the mold. Whatever you do, make sure the mold has no moisture in it: don't use wet wood - dry the mold pieces in an oven if in any doubt.

If after many months of use, the block becomes too mushroomed to suit you, melt it down and recast it.

WARNING:

Melt lead in a well ventilated area. There are lots of fumes and smoke given off, and these are best not inhaled: lead fumes are decidedly unhealthy. Also, don't stick your fingers in your mouth after using your Knockin' Block.

A CAST LEAD SHOP HAMMER

Tim sent me drawings of how he made a lead shop hammer, cast at home in a wooden mold, using copper pipe fittings as the basis.

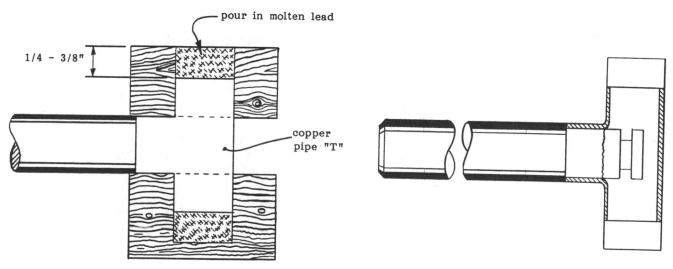

1/4 - 3/8"

pour in molten lead

copper pipe "T"

BALANCING GRINDING WHEEL FLANGES

In mid '86 I put together a short newsletter for the guys who'd bought my drawings for the TINKER Tool & Cutter Grinding Jig. That Newsletter contained ideas "my TINKER guys" had come up with in the course of building their TINKER's, as well as a few ideas of my own, and of the TINKER's inventor, N.W. Tinker. The following idea came from a TINKER builder in Philadelphia, PA, who wishes to remain nameless. I have included it here because it will be useful to most readers, not just those who have a TINKER.

The Man From Philly says: "Among other things, I found that it is important to use well-balanced grinding wheels, since any induced vibrations leave uneven marks on the tool being ground. (Getting the wheels to run true is not always easy with the stamped wheel flanges supplied with most bench grinders. GBL.) Sounds very basic, of course, but I even made special wheel flanges with adjustable weights for fine tuning. See drawings below."

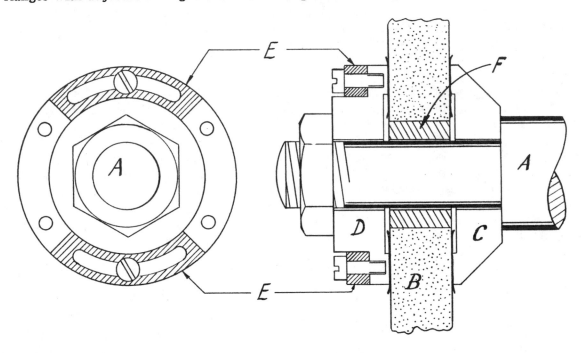

NOTE: This drawing is not to scale. It is conjectural only - I have neither tried nor seen such an arrangement myself.

The screws holding the two weights "E" onto "D" are shown with fillister heads - (easy to draw) - they would almost certainly want to be socket head cap screws, say 10-32. Flanges C & D should both be of the same diameter and usually not less than 1/3 the diameter of a new wheel. Disks of blotting paper or some similarly compressible material of a Ø larger than the Flanges should be placed between each Flange and the grinding wheel at installation.

See **Machinery's Handbook** under "Grinding Wheel Mounting" for instructions on how to balance a grinding wheel.

A properly mounted wheel should run without wobble or eccentricity. Once you have achieved this happy state and have trued the wheel with a diamond, it should run without vibration.

NOTE: it is well known that a wheel deliberately put into a slightly out-of-balance condition will cut faster and cooler when freehand grinding lathe toolbits. However, for a wheel being used for cutter grinding work, this is not considered good practice.

Also note that after the wheel has been trued with a diamond, a quick pass across the face with a dressing stick - a piece of tungsten carbide, or a "Norbide" stick (from Norton) - will open up the wheel face, letting it cut faster and cooler while leaving it nicely in balance.

AN ATTRACTIVE ETCHED FINISH FOR ALUMINUM

For an etched finish on aluminum: dip in NaOH (= sodium hydroxide) and then in a dilute solution of H_2SO_4 (= sulfuric acid) and ammonium bifluoride, to remove the resultant black smut. Looks nice. (This comes from "N.W.", a chap in California who prefers to remain nameless. From what he had to say about his work experience, I'd say he too is someone well worth listening to.

FINISHING ALUMINUM WITH A FLAP WHEEL

An aluminum oxide abrasive flap wheel is tops for a beautiful satin finish when you don't want a high polish on something. Most any of the suppliers listed in the Appendix carry these, and they are available from local sources too, auto parts stores being one place to look. (Tim Smith, again.)

A READY SOURCE FOR A LIGHT OIL, AND
A RECIPE FOR A WAY OIL
These too are from "N.W.", the chap down in California

If you want a light non-detergent oil, about like sewing machine oil, use automatic transmission fluid.

If you want a good "way lube", get some gear oil*, and add 1 oz. of MoS_2 (= molybdenum disulfide, = Molycote) and 1 oz. of STP (which is apparently a silicon based product) to 1 quart of gear oil. Shake before using. This stuff will help clean out all the little (tiny) metal pieces that can get in between moving surfaces, and keep them in suspension until they come out and can be wiped off. The "Moly" is like pouring little ball bearings all over the place - they creep into the pores in the metal, and stay there, even when the oil gets wiped up.

  *"gear oil" is not unlike ordinary engine oil, but it contains additives for extreme pressure service. Available at most any automotive service station; it is about equivalent in viscosity to a 50 weight engine oil. GBL

OIL for your LATHE CENTERS

If you want an excellent lube for running work between centers in your lathe, get some Chevron "RPM", or some "STP", or similar product. You won't need a whole can. My supply of "RPM" was drained from an empty can obtained at a service station 10 years ago. I keep it in a shallow screw-top glass jar. The stuff is so thick that when I use it, I just tilt the container about 75° and present it to the tailstock center. Once contact is made, I pull the jar back a little and the stuff behaves like thick pancake syrup - I just wrap the trailing string around the end of the center till it disengages, then replace the lid and put the jar back in its place on the shelf. This might take 10 or 15 seconds, but the lube hasn't even noticed that the jar has been standing on edge all this time, and that it had the chance to run out all over the floor. Nor has the portion on the lathe center started to drip off either.

And I hope to tell you, the stuff stands up far better than any ordinary oil would to what you know is a tough duty! Long past the point where most lubes would fail, as pressure increases as the work heats up, expands lengthwise and the centers start to scream, this stuff will be happily doing its thing. Naturally, this is not the way to carry on. One should check the tailstock from time to time during a job, and re-adjust as necessary. All I'm saying is that these hi-pressure lubes can keep you rollin' far beyond the point at which an ordinary oil would give out with a scream.

PREPARING STEEL FOR PAINTING

To prepare steel for painting, degrease it and then apply phosphoric acid - Naval Jelly. This is a mild acid, and combines with the steel to form an iron phosphate coating, which although soft itself, is impervious to oxygen (hence no further rusting) and forms a good surface to which paint will adhere, the paint then giving the final degree of protection.

A WARNING RE CADMIUM

It is well known that cadmium-bearing silver solders should be used only in well ventilated areas. But here's something you may not know: I have been told that if you even handle cadmium plated bolts (or whatever), you should wash your hands before eating! Cadmium is bad stuff - probably kill an ordinary man...

Another warning, re Cyanoacrylate Glues

"Superglue" and similar cyanoacrylate glues are useful for temporarily sticking two pieces of metal together for machining or other reasons. The adhesive bond can be broken by heating the parts. However, **this can release deadly cyanide fumes.** Therefore do it outdoors, just on the off chance you don't want to cash in your chips early. Another way to break the bond is to put the job in the deep-freeze for half an hour or so, and then give it a clout with a soft faced hammer.

A TAPPING LUBE FOR STAINLESS STEELS
David Bloom, Ann Arbor, MI

The best tapping lubricant I have found for stainless steels is from the Anchor Chemical Co., 777 Canterbury Rd., Westlake, Ohio 44145. I believe they will send a free sample on request.

A CUTTING COMPOUND IN PASTE FORM

Somewhere I saw a mention of some stuff called "Trefolex". This is a cutting compound in paste form. Put on hacksaws, taps, turned work, etc., it's supposed to be good stuff. It's made by an outfit in Britain. From the makers I got the name of their Canadian agent, and a sample of Trefolex, which I tried; I would say it's as good as heavy brown cutting oil for finish in a tapped hole, and the tap seems to me to go in a little easier - maybe. It also seems helpful on cuts with a hand hacksaw.

It cost about Cdn$25 for a 1 lb. can, which would last a lone hsm a very long time. I would be inclined to get together with about a dozen friends, buy a 4 lb can, which is about Cdn$60, hence about half the cost per lb as against the smaller size, and divy it up. The Canadian supplier is Newman Tools Inc., 3422 Notre Dame St. West, Montreal, P.Q., Canada H4C 1P2

(For exact pricing, write or phone (514) 931-2472. If you phone and are greeted in French, don't panic - they speak English too.

A HOME-MADE SUBSTITUTE FOR THE ABOVE

Bob Haralson recommends ordinary, unsalted lard, which you can get from your butcher. Get a pound or so, probably free. Take it home and warm it up on the stove until it melts, and skim off anything in the way of blood, meat, etc. Then let it cool, and take it off to your shop. On cold days, it'll be about as stiff as cold butter. On a hot summer day, it'll be more like pancake syrup, but it'll work just fine anyway.

Bob says he has tried a lot of the various potions salesmen have tried to sell him over the years, but he says he has yet to find anything that beats good old unsalted pork lard.

WHAT IS SILVER STEEL?

"Silver steel" is the British equivalent of good old drill rod. The latter can be used wherever silver steel is called for in a project of British origin. I have read that silver steel does not heat treat as well as what we know as drill rod, so be glad you can so readily get drill rod.

SHARPENING RAZORS AND OTHER FINE EDGED TOOLS

Having to shave on a regular basis is one of the afflictions of Man. Some grow beards to avoid the chore; others try every new razor that is offered. I have tried 3 types of electric razor and several of the various offerings in the safety razor category that have come along in the past 25 years or so. I mentioned to my friend Bob Haralson one day that I was shaving more and enjoying it less, or words to that effect; did he have any good ideas? Bob's response was interesting: he said he had recently had a shave from a Japanese lady barber. "...I had forgotten just how close a shave a straight razor can give. There is simply nothing that shaves like a properly sharpened straight razor."

The next morning I dug out my father-in-law's old straight razor, but when I contemplated it and my cheery little face, all freshly lathered, I chickened out. But **I knew how to do it.** How? As a child, I had many times watched my Grandfather shave, and I knew, without ever having used one myself, all the motions and facial contortions required for the upper lip, the neck, the cheek and the chin. I suppose previous generations learned the knack in like manner. But like I said, I chickened out.

> (NOTE: I have since learned that one of the keys to a comfortable safety razor shave is to wet one's whiskers **thoroughly** - not 2 or 3 swipes with a warm washcloth, but a dozen or more good wettings with HOT water. (And when I say HOT water, I mean hot enough that there is a slight odor of burning flesh when you remove your face from the cloth.) One's choice of shaving soap probably has less influence on how easy the shave will be than the time spent with the old phizog buried in a washcloth soaked in hot water.

At any rate, for those who might want to know how to put the ultimate edge on a blade of one sort or another, this article will be of interest. It will not appeal to all, but I suspect that many who read this book will find it interesting and/or useful in one way or another - there's more herein than just a few things about touching up a razor.

SHARPENING RAZORS AND OTHER FINE EDGED TOOLS
Adapted, with permission, from an article* at p104 of the
Jan 20, 1930 issue of *Model Engineer Magazine*
(*no author's name appears with the article)

A Few Remarks on Stones, Natural and Artificial

There are a great number of oilstones on the market which are all good for various purposes, but their choice and use requires not only knowledge but experience. These stones can be largely classed under two heads: natural and artificial. The principal natural stones are: -

> Turkey, coming from a small island just off Port Candia - Crete
> Charnley Forest - English
> Washita and Arkansas - American
> Belgian Hones.

Turkey stones are quite variable, some being very quick cutting and some very fine, all grades being useful for various purposes. The writer has two, one for fine work such as setting (sharpening) plane irons, and the other a coarse-cutting slip, useful for setting turning tools.

Charnley Forest stones, which vary between fine and extra fine, are similar to fine Turkey stones; they are rather slower cutting, retain their shape better, and the extra fine grades put an even finer edge on the tools and can be used for setting razors, surgical instruments, and the like.

Washita stones vary between extra quick and quick cutting; they are useful for lathe tools and as general purpose stones. They quickly wear out of shape.

Arkansas stones are fine cutting, very uniform, put a fine edge on tools, instruments, etc., and retain their shape well. About the only disadvantage is their price, but as one of these stones will last, in daily use, for something like thirty years, the annual cost is not very great. They can be used for all fine tools including razors and surgical instruments, perhaps with a final touch on a finer stone.

Belgian hones look like slate with a white material on one side. These seem to vary considerably, but all are very slow cutting and only useful to put the final touch on razors and surgical instruments. The white side is the side to use. The writer is rather doubtful if these stones are natural; they look as if the white material has been dabbed on.

The principal artificial stones are Carborundum and India oilstones. Carborundum is produced in an electric furnace, and in its raw state is a crystalline mass with iridescent sheen varying from gold to deep purple. This raw crystalline material is pounded up, graded into various sized grains, mixed with some silicaceous matter, molded into the required shape and vitrified in a suitable oven. India oil stones, the writer believes, are also a product of the electric furnace, the resulting material being called "Aloxite", which is treated in a similar manner as already described for Carborundum stones. Both can be obtained in various grades, and both are free cutting, but of the two, Carborundum is the faster cutting, and does not put such a fine edge on a tool as will an India stone, and it wears out of shape much more quickly.

The artificial stones generally are quicker cutting and the finer grades are quite suitable for sharpening the generality of tools, but for fine tools they are not equal to good natural stones of the correct grade. A Washita stone nearly equals a fine Carborundum in speed of cutting and produces a somewhat finer edge.

Another stone useful for finishing brass work and the like, but useless for sharpening tools, is Water Ayr stone, which, as its name implies is used with plain water or soapsuds. This is used for removing scratches, etc., before work is finally polished with a buff; it is also used for putting whirl marks on clock plates, etc.

Besides the stones mentioned above, the writer has seen some so-called Japanese stones, but there seems to be several varieties of these, and one that the writer came across resembled a Washita stone in quality, but not in looks, and another was like a very fine Charnley Forest stone both in looks and quality.

(The Japanese water stones of today are, I understand, considered tops for sharpening woodworking tools, and suffer only in that they tend to wear rather fast.* There have been a few advances in the past 58 years, some of them made by the Japanese. There are also man-made ceramic hones available today, which are extremely wear resistant.

*A woodworker friend of mine found that these Japanese water stones also **cut** very fast, and not just metal: he bought a water stone, took it home, and began sharpening everything in sight. Every few strokes he would pass his finger tips over the stone, probably just to feel how nice and smooth the stone was, and/or to spread some more water over it. Pretty soon his fingers began to get sore, so he stopped to have a look at them, and found that he'd sharpened the skin right off the pads of three fingers! GBL)

Lubricants, and Keeping and Dressing Stones Flat
In the writer's experience the best lubricant to use is a mixture of 2 parts paraffin (= kerosene) and 1 part each of Rangoon oil** and turpentine. This will make a stone, whether natural or artificial, cut quickly and produce a fine edge and will not gum up the stone. A Belgian hone seems to work best when lubricated with soap and water.

**a proprietary mineral oil-based English gun oil, still available today from the well known New York firm of Abercrombie & Fitch. But even if you can't readily get such fancy stuff, not to worry - read the paragraph at the top of the next page.

"The best general purpose lube for oilstones that I have ever come across is a 50/50 mix of (clean) 10-, 20-, or 30-weight engine oil and kerosene. Well mixed, and applied liberally, it helps to keep the stones nice and clean. I keep my stones in a covered pickle jar full of this mixture, so they are always clean and ready to use." (Extracted from a letter from Bill Sewell, one of my guys down in Pennsylvania. GBL)

All animal or vegetable oils should be avoided as they tend to clog the pores of the stone and make it go "sick", whereupon it loses its cutting power and the tool slides over its surface, producing a sort of metallic glaze on the stone. If this should happen to any natural stone, it can be remedied by prolonged soaking in a mixture of 3 parts paraffin (= kerosene), 1 part turpentine, and rubbing down on a flag-stone with sand and water.

The artificial stones can be revived by slowly heating to a red heat and allowing to cool off equally slowly - sudden application of heat or cold will cause them to fly to pieces, so... be warned!

All the natural stones can be rubbed down on a flag-stone with sand or coarse emery-powder and water. Carborundum stones can be dressed in the same manner. With India oilstones this is a heart-breaking performance, but a cutler's grinder will soon do the job for a few pence.

I phoned a local store that caters to serious/fanatic woodworkers, and the chap I talked to said to lap the (India) stone with a Carborundum or silicon carbide lapping compound on a piece of plate glass, using kerosene or light oil as the lubricant; alternatively use wet-or-dry paper, wet with kerosene, and placed on a hard flat surface. In either case, use a grit that is coarser than the stone to be dressed. Brownells sells silicone carbide lapping compound in various grits.

Post Script to the paragraph immediately above: From another statement I ran across since first writing the above, I gather that the surface one rubs the stone upon need not be perfectly flat - one could use a handful of sand on the concrete doorstep in front of his house, or on a cement block; there's no need to buy a $200 surface plate to rub down a $23 sharpening stone. GBL

One should try to avoid wearing the stone hollow in the first place. This can best be done by using the edge of the stone for small tools, and using one side for larger tools and the other side for plane irons, etc. If this practice is followed, the stone will seldom require truing up.

Mounting a Bench Stone
The best way to mount a stone is to make a cutout to fit the stone in each of two pieces of wood, these being alike in size, and somewhat longer and wider that the stone in question; their combined thickness should be a shade thicker than the stone. Glue a thick piece of sheet cork* on the top and bottom, respectively, of the two pieces of wood. Either side of such a box can be opened for use, and the cork base prevents the stone slipping on the bench.

*Where to get sheet cork? Look in your Yellow Pages under "Cork". or: Buy and rip up a cork bulletin board. or: Phone a seller of flooring materials. GBL

Sharpening, Honing and Stropping a Straight Razor
The question of the correct angle at which to grind and sharpen tools is beyond the scope of this article: suffice is to say that the angle should be kept constant and a rounded edge avoided. As to the angle to employ for straight razors, more below.

The writer does not know if it is a peculiarity personal to him or if others are afflicted with the same trouble, namely that of having a very tender skin and very tough whiskers. In his own case, this unhappy combination renders the use of a safety razor about as futile and un-comfortable as trying to shave with a garden rake - the (safety) razor simply hops over the bristles and takes out a scoop of skin in between.

The difficulty in using a straight razor is to sharpen it in the first instance, and to keep it sharp afterwards. But a straight razor, if you know how to use it and how to keep it sharp, is superior to any other type for a smooth shave. (Which is exactly what Bob said. GBL)

The first thing to do is to get a good razor, or a pair if possible. If they are not in good order as bought, they must then be "set", which is to say, properly hand ground by the new owner.

The writer has found that a slip of lithographic stone (which has no cutting action in itself), to which a mixture of Tripoli powder in turpentine has been applied, makes a perfect razor hone.

The razor should be held dead flat on the stone so that the back and cutting edge bear equally. Press not too hard, but at the same time firmly. A reciprocating rotary motion is then given five or six times on one side of the blade and then on the other. When the edge appears sharp, the setting can be finished off by drawing the razor over the hone towards the cutting edge, at the same time drawing the razor off the stone from heel to tip, alternately on either side of the blade. The front edge of the stone should be slightly rounded off so as to prevent the possibility of the razor digging into it and spoiling the edge.

If lithographic stone is not available, a bit of flat iron, say 6" x 1-1/2" x 1/8", with one side thickly tinned with solder all over, the latter then being filed and scraped quite flat, answers as well if not better than the stone.

If the razor is in a bad state, "Diamantine*", obtainable from any of the watchmakers' tool stores, which are as thick as plums in Clerkenwell Road**, acts as a quicker cutting agent than the Tripoli powder, which by the way, is obtainable from the same source.

*"Diamantine" is an extremely fine alundum (= aluminum oxide) abrasive product from the Rheinfelden area of Germany. It is, I am told, sold in three grits, fine, medium, and coarse; all are exceedingly fine. Diamantine is considered **the** thing to use to get the highest possible polish on steel. For a mail order source, see end of this article. GBL.

The reference to Clerkenwell Road will mean little to most readers of this book. Clerkenwell Road is a street in London, England that was at one time, and still is, to a lesser degree, **the place to buy clockmakers' supplies and all else related thereto. GBL

The writer set his razors once during the Great War, and one of them once since, after someone borrowed it.

The reason razors require resetting so often in most users' hands is due to a wrong system of stropping. This is illustrated in Fig. 1, to an exaggerated extent. As will be seen, the strop is bent down by the pressure applied to the razor blade and passes up over the edge of the blade at a more or less acute angle. In the illustration a strop of the leather strap type is shown, but a stick strop has the same effect if too much pressure be applied, or if the leather surface is too springy, or a combination of the two.

The writer uses a (rigid) wooden strop, one side being covered with end-grain wood and the other with sole leather.

(Now pay attention, guys, 'cause this is where the writer of this piece really brings out some interesting info. I'll interrupt in a few spots to interpret. GBL)

The best wood found so far is Ayous white-wood. (= African white-wood; Obechi wood; GBL) Any of the bastard mahoganies, such as "Butternut" (technically, the equivalent of American black walnut; GBL), can be used. What is wanted is a hard wood with an open porous grain capable of retaining the dressing without springing under the edge of the razor. A strip of wood about 3/16" thick is sawn off the end of a 1-1/4" plank, and pinned with pegs cut from the same wood and glued to the board from which the razor strop is to be made.

(The board is hereafter referred to as "the stock". Obviously, from what little the writer says - or rather, doesn't say - it should be about 1-1/4" wide and I would suppose 12 to 15" long, and maybe 1/2 to 5/8" thick. Also, with the adhesives available today, I suspect one could skip the pegs. GBL)

Bad practice:

Fig. 1

Leather strop

A piece of the best-quality heavy sole leather, obtained from a shoe repair shop, should be soaked in water and dried fairly close to a fire to harden it and then the smooth (hair) side glued to the stock. When dry, both sides are planed down, the end-grain wood until dead flat and true, the leather side until all the soft pithy part is removed.

(I suspect there is a lot left unsaid in that sentence about the sharpness of the plane blade needed to plane end grain walnut! But if the stock were good and flat to begin with, a couple of passes through a thickness planner - or through a table saw with a carbide blade - would do the trick on the wooden side, and a good sharp hand plane would do fine on the sole leather side; I would be inclined to get the wood side copacetic before gluing the leather to the other side. GBL)

The wooden side is dressed with a paste made of washed (= "levigated") Tripoli powder mixed with 1 part beeswax to 4 parts castor oil. (Brownells sells "Tripoli polish", which is probably not the exact same thing, but is the closest I can spot right off the bat. GBL.)

The leather side is treated with carbon obtained from charred straw, mixed with the same oil/beeswax recipe. The straw should be washed thoroughly to remove any grit, dried, stuffed into a cocoa tin and placed in a hot fire, a small hole or two being punched in the lid to allow the gas to escape. The writer believes rye straw is the best, but any ordinary straw seems to answer quite well.

All grasses have a very thin coating of silica on their stems; when the straw is turned to charcoal, the silica is separated in very fine flakes which form an exceedingly fine abrasive. The castor oil and wax keep the leather soft and form a semi-adhesive paste for the cutting agents.

The writer finds that it is only necessary to use the wooden side about once every two weeks; the leather side is used every day. The Tripoli requires renewal about once a year, the black paste once a month.

When the wooden side is used, care must be taken to wipe off any Tripoli paste adhering to the razor before going on to the leather side, because if any Tripoli gets on the leather it will drag the edge off the razor.

The writer could never understand why extremely hollow-ground razors were supposed to be superior. The cutting edge is the important point, and the angle of this is governed by the thickness of the back in relation to the width of the blade. These two dimensions fix the angle at which the edge is set and stropped (see Fig. 2).

Fig. 2

A   B

The flexibility of the edge may be advantageous in that if too much pressure is employed in either of these sharpening processes the blade will bend up to adjust itself to the angle of the slip or strop. In the writer's hands a slightly hollow razor is just as good as an extra hollow-ground specimen, the two being used alternately without any appreciable difference in cutting ability being noted; in fact, sometimes one seems to work best, sometimes the other, but neither is ever far from perfection.

NOTE: Readers in larger cities should be able to find a local source of such things as Diamantine powder, by looking in the Yellow Pages under "Watchmaker's Equipment & Supplies". If this bears no fruit, you might want to contact the undernoted mail order outfit, which is well known within the watchmakers' trade. My experience of them is nil; I simply got their name from a like outfit here in Vancouver. Their catalog can be had at no charge upon request. If you can't find something of interest in it, I'd be mighty surprized. Address: S. LaRose, Inc., 234 Commerce Place, Greensboro, N.C. 27420.

COPPER VISE JAW LINERS

Where can you readily get a piece of sheet copper from which to make soft jaws for your bench vise? Easy: Cut yourself off 2 pieces of 1/2" copper water pipe (super easy to get), each about 3/4" longer than your vise is wide. Split each piece lengthwise with a hacksaw. Pry open with pliers, flatten, put in vise and bend to shape, cut a notch at corners, and fold at ends - Voila! - not-falling-out vise jaw liners! The copper will work harden while you are beating it flat. To anneal it, heat red hot or nearly so with a propane torch, dunk in cold water, and carry on clouting.

HONING LATHE TOOLS

A 6" India aluminum oxide abrasive stick of triangular section is a handy tool for stoning lathe tools. Bill Fenton gave me the one I have, and I think it is a "medium" grit; in any case, that would be the right grit to choose if buying new for this purpose. This stone cuts surprisingly fast on HSS.

When touching up a lathe tool (or the cutter in a fly-cutting head for the mill) I usually take the toolholder/toolbit off the Quick Change Toolpost and hold it in my left hand at the edge of my masonite-topped workbench. I apply the stone like a file, steadying it by contact with the edge of the bench as well as the toolbit. This helps produce a facet, rather than a rounded surface.

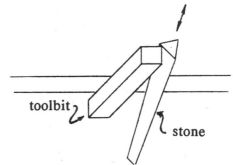

A word of warning: Do not drop such stones - they break easily if dropped on the floor. They rarely wear out, but are often dropped, so Norton sells lots of them in spite of their long wear life. Where to buy? Brownells sells 'em, and outfits like MSC Ind. Supply Co., J&L Sales, and your local Norton distributor.

"MF" Aluminum Welding Rod

Generally, the welding of aluminum is a tricky proposition because the stuff gives you no warning that it is getting hot until all of a sudden it melts. If you want a humbling and/or frustrating experience, try welding irrigation pipe!

I have been told flat out absolutely "You can't solder aluminum!" - this from a master sheet metal worker in his eighties. When you see what he does with sheet metal (and many other materials) you have to figure he knows what he's talking about.

Well, one day along comes a package in the mail from Tim Smith. Inside were 3 or 4 sticks of what looked like aluminum wire. Accompanying same was a xerox of some maker's literature about this aluminum welding rod, and a letter from Tim raving about what great stuff it is. He says it makes joints stronger than the base metal, using nothing more than a propane torch. He says he makes all kinds of little brackets and stuff wherever he has occasion to want to weld aluminum. The stuff also works on "white metal", e.g. steering wheel horn rings, window cranks, etc. Just preheat and tin the metal per instructions.

Post Script to the above: In early August '87 I attended an antique machinery show in Lynden, Washington with Bob Haralson and a couple of other like-minded friends. There was a chap there selling this same stuff. He had, among numerous other items, half of a chainsaw outer casing, which had been "cracked" for demo purposes by running it into a bandsaw for about 6". The "crack" had then been repaired with a bead of this stuff run along the outside of the housing. The bead had not been ground down flush but it had been smoothed up somewhat with abrasive paper.

We turned the housing over and noted that the back side of the cut was not filled in at all. Jake said it looked strong enough, but the penetration was poor... "...But I have used that stuff myself, and it is good. Whoever did this just didn't use enough heat. I've done repairs like that, and got full penetration, just using a propane torch." Jake Wiebe is a skilled metalworker, and if he says the stuff is good, and Tim says the same thing, that's plenty good enough for me. Aside from the trade name given in the caption, I believe it is also known as "Lumiweld".

RUBBERDRAULICS
Hal Robinson, Arlington, MA.

Hal was one of two readers to suggest the use of silicone rubber instead of oil as a means of transmitting pressure to the little plunger behind the graduated slip ring, in the Graduated Lathe Leadscrew Handwheel project at page 71 of **TMBR**. (My apologies to the other chap who suggested this idea - I have lost track of who he was.)

Hal says: "I have used cast-in-place silicone rubber instead of hydraulic oil when a movement of only a few thousandths is required. Short pistons at both ends of the chamber require less accuracy if rubber is used instead of oil. There are no leaks, and no need for O-rings."

This is a first class idea for this particular job, and in other applications where very little movement is required. I'd call it a "rubberdraulic" fitting.

(A note about this idea was added to **TMBR** in the second printing thereof, so most readers will see it there, but for the benefit of those who have a first edition copy of **TMBR**, I have put it in here too. GBL)

THE CROSS FILING TECHNIQUE
More from Hal Robinson, Arlington, MA.

I do a lot of fine file work in the lathe, on soft and gooey metals - silver, copper and aluminum. I use lots of oil, and frequently brush out the file with a fine file card. I also use a sewing machine needle in a pinholder handle to run down the grooves of files if the clogged metal is particularly difficult to get out. This can be time consuming but I enjoy it as a little break in the routine. I sit back with my feet up and pretty soon I have what looks like a brand new file.

Also: use a piece of sheet brass (or copper) to push across the face of a file, bulldozing out any material clogging the teeth. The use of chalk is as old as the hills, yet it is surprising how many - even old hands - have never heard of it.

I have been using what I call a criss-cross filing technique for years, for all the good reasons you mentioned on page 8 of **TMBR**, and two others which I think are important.

  1. Successive strokes of a file in one direction tends to encourage chatter to build up whether you are working on the lathe or at the vise. Changing the filing direction eliminates this annoying problem.

2. Also, files tend not to clog up as quickly when one uses a criss-cross technique. The "chips" taken by each file tooth are intermittent and tend to be granular rather than long curls or large build-ups, and so they fall away more easily.

OIL AND STEEL WOOL FOR NICE FINISH

Another good way to get a nice finish on parts being turned in the lathe: Polish with fine steel wool and oil. (Bill Fenton again.)

Also, if you have a piece of blued (or bright) steel that has become somewhat rusted, try rubbing the area with steel wool and oil. I have used this myself on hot rolled sheet steel, with pretty good results, and again on a rifle barrel. Not a complete cure, of course, but sometimes it'll do a very good job, particularly if the rust is only superficial. See also Gunsmith Kinks (both Vols.) for more ideas along these lines.

1200 GRIT ABRASIVE PAPER

Did you know there is such a thing as 1200-grit wet and dry paper? Tim Smith sent me a sample. If you thought 400-grit ditto was fine, wait till you try this stuff - it's like rubbing a mink coat.... well, not quite. You can get it at any store that sells auto body supplies.

PAINT STICKS AND SANDPAPER
Ted Lusch, Mt. Vernon, WA

And another one, related to the above: Get some paint stirring sticks at a paint store. Spray with contact cement from a spray can and stick abrasive paper thereto. The use of spray-on cement avoids lumps under the abrasive paper. Use the resultant abrasive stick like a file.

(The wooden paint stirring stick may be a thing of the past: the last one I got was made of plastic, had large holes in it and raised edges. If you have a woodworker/cabinet maker friend, ask him to make you some thin sticks from scrap wood by running same through his thickness planer. GBL)

SOME FLUX REMOVAL TIPS
Adapted, with permission, from a "Shop Talk" by
Lou Kashycke, *Live Steam Magazine*, November 1978

To remove all traces of acid flux after silver soldering operations, wash the job thoroughly in hot soapy water. Lux soap is good. Result is no further corrosion. Finally, rinse well in warm water to remove soap or curds formed by interaction of soap and acid.

To remove silver solder flux: Quick, and safer than hot water, sulfuric acid, and/or chipping and scraping, is the use of one tablespoon of baking soda in an 8 oz. glass of water, plus an old toothbrush to scrub with.

FLOWER ARRANGING ROD AS A WELDING ROD
With permission, from an article on steel sculpture
by Don Greytak, *Live Steam Magazine*, October 1978

Flower arranging rod is mild steel, and can be used as a welding rod. Welds do not harden even if cooled very quickly, whereas this can be a problem in welds made with gas welding rod.

SETTING WORK FLUSH WITH TOP OF VISE JAWS

To get a piece of work flush with the top of the vise jaws in your vertical milling machine, tighten the vise while pulling the part up against a lathe toolbit, a burr-free piece of CRS, or other flat surface placed face down on top of the vise jaws.

THREE TAPPING HINTS
With permission, from a "Shop Talk" by
James Hamill, *Live Steam Magazine*, October 1978

1) Discard (or sharpen) dull taps. Sharp taps show sharp, bright edges when viewed with a magnifying glass. (Better to throw them away than to have them break in a hole.)

2) Run a smaller tap of the same pitch into a hole before the full size tap. e.g. 5-40 before 6-40. or 4-40 before 5-40, or 6-32 before 8-32, etc.

3) Have two sets of taps - use the best (newest) for the most valuable jobs.

FAST REMOVAL OF TAP CUTTINGS

Hal Robinson again: "I consider an air compressor an absolute necessity in the shop. I flared a 3" length of 0.050" stainless hypodermic tubing at the back end, and epoxied it into the screw-on nose piece from inside, on one of my quick connect handles. The result is an ideal nozzle for blowing chips and oil out of deep tapped holes.

CAUTIONS:

a) chips come out of the hole like little bullets, so close your eyes or take other precautions.

b) any screw-on end on an air hose handle can come unscrewed, so MUST be checked periodically. This one, with a thin steel tube protruding from the business end, would be particularly dangerous if propelled with 90 psi air." (Dangerous? It might as well be a blow gun dart! GBL)

Comment: Compressed air is great stuff if used with intelligence. My friend Bob Haralson would not have an air compressor in the shops he ran: it is almost impossible to prevent men from cleaning their machine tools with compressed air if it is available, and this practice drives small metal cuttings into areas where they are least wanted. Even if you fire a man for doing this, the damage is already done before he is out the door. GBL

PRODUCING A BLIND OPENING
BY CHAIN DRILLING

The idea of stock removal by chain drilling will not be news to anyone reading this book. However, some of the points noted below will be useful to keep in mind for such occasions.

1. Lay out the line you want to cut to, and the line the hole centers will follow.

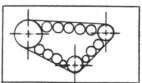

2. Make a punch of some sort to space out the hole centers the correct distance apart. One way to do this is to weld two center punches together - it's crude to look at, but it'll work.

3. If round corners are desired in the finished profile, drill/ream these next.

4. Generally, if chain drilled holes are sized to leave a web of solid metal 15 to 30 thou thick between adjacent holes, a flat ended hand broach, which can be forged up from a piece of hot rolled tool steel as shown at right, will be found to be both a quick and safe tool to remove this web.

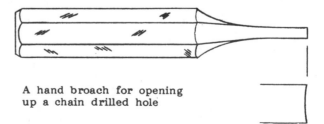

A hand broach for opening up a chain drilled hole

5. Alternatively, the holes may be spaced and sized so they actually touch. This risks drill

breakage. A better practice is to drill first with an undersize drill, and then drill out these holes to a size that will cause adjacent holes to touch.

6. File or mill the opening to finished profile.

QUICKIE BANDSAW BLADE WELDING JIG
Tim Smith

To hold a bandsaw blade for welding, put a cotter pin over each end of the blade, and grab the two cotter pins in your bench vise.

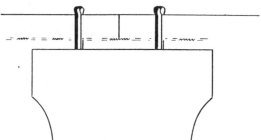

MODIFIED OIL CAN SPOUT

Tim modified his oil can spout as shown below so he could get Gitts oil cup lids open using the spout of the oil can.

TUBE FLARING IDEA

Tim again: Here's a simple tool to put a flare in small thin tubing if ever required.

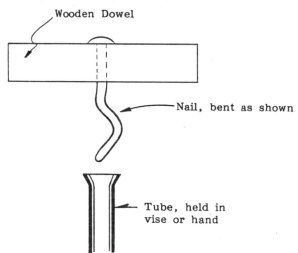

Wooden Dowel

Nail, bent as shown

Tube, held in vise or hand

Addition to the second and subsequent printings of TMBR#2:

On December 8/88, I had a call from Cliff LaBounty of Maple Falls, Washington. Cliff specializes in re-boring and re-rifling rifle barrels. He liked The Machinist's Second Bedside Reader just fine, but he drew my attention to an error in The Bullseye Mixture, the fiction story which starts at page 163 herein.

Cliff pointed out that one cannot simply machine down the shank of a Springfield barrel, re-thread it to suit a Sharps Borchardt action, poke a .30/40 Krag chambering reamer into the existing .30/06 chamber, and thereby produce a proper .30/40 Krag chamber. The dimensions of the .30/06 chamber are larger in most particulars than the .30/40 Krag, such that the .30/40 reamer would cut almost nowhere except in the neck area of the existing chamber. This was an oversight on my part - I should have looked up the dimensional specs on both cases.

I want to point out this error to all who read The Bullseye Mixture, simply to be sure that no one comes to grief through trying to emulate my main character's re-chambering activities. Chalk it up to "artistic License", if you will, and enjoy what else the story has to offer, recognizing that it was not written to provide instruction on gunsmithing, but rather to consolidate and present, in a most palatable form, full and detailed info on old-time pack casehardening methods. To the very best of my knowledge, that material is NOT flawed.

SUBRECKY'S GADGETS - TOOLS YOU CAN'T BUY,
or... New tools for the Oddleg Artist,
plus some tricks to keep your tail outta the cracks

I received a package in the mail one day from Jim Subrecky, an electronics technician at a radio telescope observatory in northern California. In the package was a layout tool you won't find in any catalog, plus drawings for another device similarly unavailable commercially. Jim said both of these items had been constructed during discussions with Bob Millray, of Saugus, CA. to solve problems associated with some work they were doing, and he sent the info about them along for my edification and possible inclusion in **TMBR#2**.

Jim views Bob Millray kinda like a cross between a wizard, a genius, and the best teacher a fella could ever hope to have, and emphasized that credit should be given to Bob for these ideas. Jim is one smart dude himself, as you will see.

Let's look at the layout tool first. Jim says: "I used to work at a place where we made prototype cameras, and we often had to mill intricate detail in sheet metal (i.e. rotary switch contacts). Many times in such work it is helpful to lay out scribe lines before milling with a cutter. This way, you can often see where you are about to make a mistake. This tool, which is very cheap and quick to make, puts a relatively constant force on the point. The pressure can be adjusted somewhat by the allen set screw that holds the spring captive.

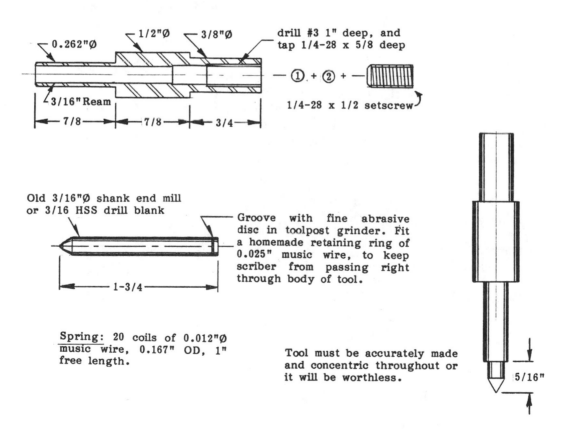

"The tool is gripped in a collet in the mill. The spindle (power off, throughout) is lowered enough to put a light pressure on the workpiece via the internal spring. You then scribe the part using the dials on the mill. This offers a visual check of what will happen when you mill the part, plus it leaves a scribed line which gives a second visual check for when you subsequently do the actual machining of the part.

"Sometimes you may have a part that is very expensive or has a lot of time invested in it, and you cannot afford to make a mistake. This is one place where you would use this layout tool."

(Another use I can visualize for this device is for doing an entire scribed layout on the face of a job, using the mill and its table feeds as a layout machine. GBL)

"In making* this device, care must be taken to achieve high concentricities and close fits so that there is very little side play on the point shaft; if these conditions are met, the tool is a winner! Once you use it, you will be hooked."

*Jim said to make this item from 1/2"Ø CRS. The one he sent me was so made, and it sure works nice, but I would be inclined to make it from say 5/8"Ø CRS, thus: Start by facing both ends, to clean them up and bring the material to length. Then drill/ream the piece 3/16"Ø full length. Loctite the job onto a piece of 3/16 drill rod just slightly longer, which has been accurately center drilled at both ends. Then turn the working OD to spec. Finally, chuck by the OD, and finish all other work on the job. GBL

A Multi-Diameter Edge Finder Adaptor

Ever get tired of changing collets when you need to use an edge finder? I have two of these, one with 3/16, 3/8, & 1/2" OD's, and one as shown in the drwg at right. You can't buy these, either, but they're easy to make. If you don't feel like doing a press fit, use Loctite.

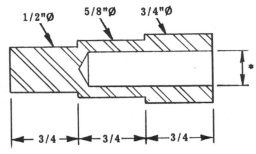

*Bore 0.001" under diameter of edge finder; press edge finder into place.

A Slitting Saw Arbor

If you measure a sampling of slitting saws, you will find they are "right on" for the given hole diameter. The thing to do if you want these saws to run nicely is to make a very carefully fitted arbor, giving the business end of the arbor one or two tenths clearance only. The design shown here also allows you to cut very close to the bottom of an inside corner, plus it allows you to change saws without removing the arbor from the mill (or lathe).

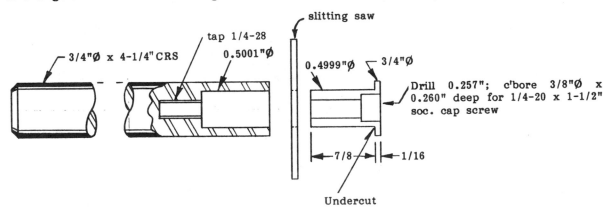

The Seven Baggies of Subrecky

Jim also sent me another package containing 7 plastic baggies, plus a note saying, "Herewith my collection of tooling and workpiece samples which I collected while working with my mentor, Bob Millray. All of them are the products of his genius."

In each of the baggies was a special tool made to do a specific job, plus a sample of the workpiece produced therefrom. An explanatory letter accompanied the baggies. Every one was interesting - sure as anything one or another of them will be useful to you somewhere along the way. I made drawings to illustrate each of them, and the following paragraphs are quotes from Jim's letter describing them.

What was in the Baggies

Baggie #1. Problem: We encountered an impeller bushing which had been pressed into a 1.750" diameter hole in a turbine-type water pump. There were no protruding edges to catch, and we had to remove it. A similar problem would be removing valve seats from a VW head.

Solution: Using a heliarc, Bob formed a puddle on the inside circumference, and "walked" this puddle right around the interior of the bush. As the puddle solidified, it shrank the bush, which then just fell out of its hole. If you measure the O.D. of this ring you will see just how much it shrank. (It is now about 1.742"Ø. GBL)

Baggie #2: Problem: How to hold thin, odd-sized diameter parts in a collet with "reasonable" repeat positioning.

Solution: Make a sleeve as seen in Fig. 1, and size it to fit nicely in a standard collet - in this case a 1" 5C collet. The internal and external seating steps guarantee minimal run out. Use a single slit to guarantee a concentric holding of part. Sleeve seats against collet nose to give "pretty good" repositioning of part after removal. It does not give "excellent" repositioning, because like any part held in a standard 5C collet, the part's position depends on the collet's taper. If you reinsert the part and put just a little more or less force on the collet closer, the part will be drawn in deeper or less deep.

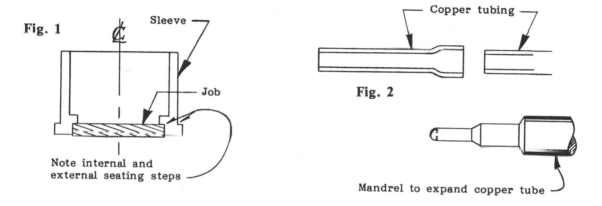

Baggie #3: Problem: How to expand the end of a piece of copper tubing so two pieces can be sleeved together.

Solution: Make a mandrel as shown (Fig. 2.), with a good finish on the working portion. Hold the tubing in a collet in the lathe headstock. Put mandrel in the tailstock. Oil mandrel heavily. Run the lathe at a slow speed. Slowly advance mandrel via tailstock. If the copper tubing work hardens, anneal and carry on.

> TIP: A solution of ivory soap in water is a good lube for tube expanding and deep sheet metal drawing operations, e.g. the stamping of something like say an engine oil pan. One might find the same soap 'n water mix good to use in connection with the above idea.

Baggies 4 and 5. Problem: How to punch out shapes as in Figs. 3 and 4 in thin sheet metal.

Solution: These punches were all made on a standard Bridgeport mill. The dies were held in the vise and the punches were held in the spindle. You may question that the material is stainless on one part, while the die is soft brass. We only needed one good part off the die, and this did the job. The items in Fig. 3 were for a ratchet mechanism, while the piece in Fig. 4 was used to hook the end of a small spring to.

Baggie #6. Problem: We had to make a very small solenoid, which requires that the metal of the casing and the core be continuous.

Solution: A cutting tool, as at Fig. 5, to use on a mill or lathe. Bob Millray made some of these cutting tools with wall diameters down to 0.010".

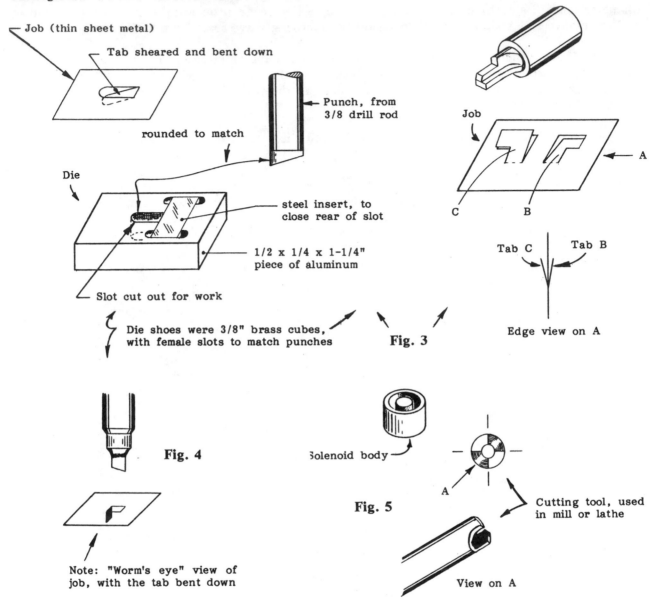

Job (thin sheet metal)

Tab sheared and bent down

Punch, from 3/8 drill rod

rounded to match

Die

steel insert, to close rear of slot

1/2 x 1/4 x 1-1/4" piece of aluminum

Slot cut out for work

Die shoes were 3/8" brass cubes, with female slots to match punches

Fig. 3

Job

A

C   B

Tab C   Tab B

Edge view on A

Fig. 4

Note: "Worm's eye" view of job, with the tab bent down

Solenoid body

A

Cutting tool, used in mill or lathe

Fig. 5

View on A

Baggie #7. Problem: Turning the O.D. of parts of very thin sheet material on a lathe.

Solution: Sandwich the material between two pieces of bulky material, as at Fig. 6 at right. This came up when we had to cut some thin printed circuit boards to make wiper switches. The large end is held by a collet, and a live center presses against the small end. A cut is made through plastic and all. As you would say, Guy, "...it worked like a charm." *(Do I say that?)*

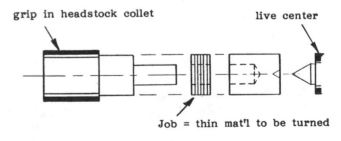

grip in headstock collet

live center

Job = thin mat'l to be turned

Turning OD of parts of thin sheet material on a lathe

Fig. 6

130

(I like the punches for one-off stamping jobs. On a much cruder level, but in the same vein, just a while ago I got tired of removing the lid of the jar which serves as a "Piggy Bank" for our US coins, every time I wanted to put some coins in it. I figured that if a machinist couldn't produce a creditable Official Piggy Bank slot, he had a pretty serious problem. So I got busy with scriber and steel rule, and laid out a coin slot to the standard coin slot dimensions of the American Piggy Bank Manufaturers Association. I put a hardwood block in the bench vise, and set the jar lid on top of same. Using a small hammer, I then tapped a lathe toolbit through the lid all along the layout line, shearing out the material I wanted to remove. It worked like... ummm.. well, yeah... it worked like a charm, and took only a few taps with the hammer to flatten out the edges of the metal after making the cut. I touched up the edges with a needle file, and it was done, nice as need be. GBL)

WHAT IS THE PURPOSE OF THE BALL END
ON A SURFACE GAGE SPINDLE?

Hal Robinson also raised an interesting question: What is the purpose of the ball on the end of the spindle of a surface gage?

I always supposed it was just a nice finishing touch, but after the question was raised I had occasion to phone the Starrett factory in Athol, and I asked my contact there if he knew. Here is the answer, and it cost me a nickel to find out, so pay attention:

"The groove that is developed behind the ball serves as a safe place to park the curved end of the scriber when the surface gage is set away in one's tool box."

I think this is an interesting point - no pun intended. I mentioned it to Frank Stubbs, an English model engineer who visited me one day. He told me that when he was a young man he had worked in a big machine shop; if you drew a surface gage from the tool crib and returned it other than with the point tucked in as above, you would hear about it in no uncertain terms from the tool crib foreman!

A USE FOR WORN OR BROKEN HACKSAW BLADES
Fred Whaley, Camano, WA

Concerning what follows, Fred says his experience is limited to the all-hard high speed steel variety of hand hacksaw blade.

"....Their conversion to knife blades is too mundane to mention but I will anyway. It came about when I broke a new blade through carelessness. To ease my guilty feelings I had to make something useful of the evidence. A paring knife of 2-1/2" blade length was the result. While not corrosion-resistant in the stainless sense, it does not rust in the dishwasher environment. It remains sharp three times as long as commercial paring knives. Even out of its element, our shop cutting tool material is certainly entitled to respect.

"In the shop, use the HSS blades as welding filler rod. Build up a deposit on mild steel stock, and then grind into cutting edges for lathe bits and boring bars. Keep the torch flame adjusted on the carburizing side to avoid oxidizing the chromium content of the HSS. Chrome oxide forms an infusible slag entrained in the deposit. In my experience, no further heat treatment is necessary."

Rubbing Away
Brian King, Tustin, CA

"Craytex" products are an abrasive in a rubber matrix or carrier. They can be had in the form of blocks of material (like an oversized eraser) for hand use, or as mounted points for use in a die grinder or similar. While he has had little joy from the mounted points, used in a Dremel tool, Brian says he has found that he can get a very nice finish on work running in the lathe by applying a Craytex block to the underside of the work, so that rotation of the work wants to pull the Craytex block away from him, rather than shoving it back at him. (This is a safety thing, I think - the finish produced wouldn't be any different one way or the other. GBL)

As a logical lead-on from the above, suppose you have an awkwardly shaped job to file in the lathe, and you figure the danger of the file being shoved back at you is greater than usual. (In fact this is an ever-present danger when filing in the lathe, and the consequences can be mighty serious if the file lacks a handle, and should happen to be caught by one of the chuck jaws.) In such a case, reverse the file, hold it by the tip, and present it to the revolving work from the underside. If anything grabs the file, it will be pulled from your grasp, not driven back into your hand. And watch you shirt sleeves/cuffs when doing all such things. GBL

HIGH FINISH, HIGH PRECISION HOLES

There is a simple technique used to bring a machined hole to a much higher level of roundness, size tolerance, surface finish and surface durability. The technique is called "ballizing", or "ball-sizing", and is fast, simple, and well developed, although not widely known. It involves nothing more than forcing a hard, wear-resistant ball through the hole to be ballized. The ball must be slightly larger than the as-machined hole, and the ballized hole will end up slightly under ball size - some springback occurs behind the ball as it passes through the hole.

The ball can be forced through the hole with an arbor press, or with hydraulic pressure. Blind holes can be handled by attaching a push rod to the back side of the ball, so that it can be pulled back out of the hole. Soldered joints in copper tubing etc. are sometimes ballized to iron out the solder bead, to minimize disruption of the flow of fluid in the finished line.

As I said above, the practice of ballizing is well developed; at least one firm supplies tungsten carbide balls and offers its accumulated knowhow for implementation of the technique. The company is Spheric Inc., 9450 7th St., Ste. H, Rancho Cucamonga, CA 91730 (1-800-824-2099).

Footnote for the gunsmith/shooter/machinist: think about making bullet sizing dies for cast bullets by ballizing a reamed hole to size. The trick would be to know what size ball to use to make a hole of a given size - this is the sort of thing Spheric could probably tell you quite readily. Ballizing would probably not be the way to go if you wanted to make one or two bullet sizing dies of a given size, but if you wanted to make several, it'd sure beat lapping 'em to size!

━━O

A Handy Tool Tray
Wayne Griffith, master clockmaker,
Murfreesboro, Tennessee

If you want a nice neat tool tray, get thee to an office supply store and buy a small, framed, cork-faced notice board. Spray an adhesive on the cork surface, and glue on a piece of green felt. Attach three pieces of wood to the back side, the two outer ones being a little deeper than the middle one. The one in the middle can be grabbed in the bench vice, or if you want to set it directly on the work bench, the deeper outer legs serve as feet.

Wayne tells me there is an American magazine which caters to the clock repair trade, i.e. the guys who make their living getting their hands dirty fixing clocks, as opposed to the hobby clock makers and clock collectors. The magazine is called "Horological Times," and is available from:

> The American Watchmakers' Institute,
> 3700 Harrison Avenue,
> Cincinnati, OH 45211

Wayne has made two clock repair videos, available from S. LaRose, Inc., a supplier mentioned in the section on sharpening razors and other fine edged tools.

132

IF YOU'RE LOOKING FOR A DRAFTING MACHINE...

Some guys work without drawings - how they do it is beyond me. Other guys can't read a drawing. (Or they say they can't. Some drawings are lousy, no question about it, but assuming the drawing is properly done, I suspect that the guy who says he can't read it is in fact unwilling to take the time to study it so that it becomes clear.)

I regard drafting as an integral and enjoyable part of my shop activities. One can easily spend too much time working up drawings far nicer than needed, but if you like doin' it that way, who's to stop you?

One thing that can speed up drafting operations is the use of drafting templates. I have several templates which I find useful: two of them produce circles from under 0.10" up to about 3"Ø, another does hexagons (for hex sockets, nuts, bolt heads etc.), plus another for ellipses. The most recent acquisition was a set of three for doing Roman upper case letters and numbers in 1/4, 3/8 and 1/2" heights - these were something I had not seen before, and are put out by Pickett (See Appendix); they produce a nice result. The cost of some of these templates may shock you when you first price them, but they earn their keep, and you will not regret buying them.

The biggest aid to drafting is a good drafting machine. I have an arm-and-elbow type drafting machine on which I do my drafting, including all the illustrations for this book.

This drafting machine was given to me nearly 10 years ago by my friend Archie McDougall, who was a consulting civil engineer, and one of the smartest men I've ever worked with. He had several old drafting machines stuffed in a corner in his office, machines that had been obsoleted in an interesting way:

Whenever Archie hired a new engineer or draftsman, he would assign him to one of the several drafting tables in the office, and then look at the drafting machine attached thereto disapprovingly. "Och noo, that machine looks pretty tatty. Just you trot over tae that drafting supply outfit doon the street, pick yourself out a new machine, and tell 'em tae put it on our account." (Now, it does not take a genius to see the psychology behind this: The new guy has just been told to go and pick himself out a brand new toy, because this old one is not good enough for him. Is he going to like working for Archie? I hope to tell you, we loved him.)

Archie would put the "tatty" machine in the corner with the others already there, or move it to his own table, and use it himself.

Knowing that he had several of these old machines laying about, I asked him if he would sell me one of them. A day or two later he phoned and asked me to come up to his house, as he had something for me. When I got there he handed me a fine VEMCO drafting machine, and wouldn't take a penny for it. "Och aye, it's moore trouble tae find out what it's wurrth than tae just gi'e it tae you."

I took it to the local VEMCO dealer and had it serviced, bought a pair of oddball scales for it at half price, and shortened the two tubular arms by about 6" in my lathe, which necessitated that the driving bands be cut, shortened, and re-riveted. I made a 21"x30" drafting board from hi-density particle board, put a piece of green drafting surface on it, and for $70 total I had a portable drafting outfit which would have cost me about $450 new. I can set it on top of my work desk, and equally quickly set it aside when not required.

(The above account of how I came to acquire my drafting machine may give you a clue as to how you might be able to buy a good used drafting machine at a reasonable price. Look in the Yellow Pages under "Engineers, Consulting" and start phoning. This is all the more pertinent today because of the rapid advent of computer assisted drafting, with the result that many conventional drafting machines are being dumped.)

A REPLICA LUNKENHEIMER WHISTLE

I visited a chap hereabouts one day who is a self-confessed whistle nut. We'll call him Mike. Mike has a collection of whistles off steam tugs, factories, sawmills, etc. - maybe 30 in all, from 1/2" dia. x 2 1/2" tall to 8"∅ x 3' tall. All were in beautiful condition, and fully functional.

One of the smaller whistles in the collection caught my eye. It was a Lunkenheimer. Mike let me give it a once over with calipers etc., and I give below working drawings.

"What's a Lunkenheimer?" you ask. A Lunkenheimer whistle, often abbreviated to just "Lunk", is a product of the Lunkenheimer Company, of Cincinnati, Ohio; Lunkenheimer used to manufacture high quality steam whistles, but no longer does, as far as I know.

Apparently steam produces a nicer note from a given whistle than will air. For sure the note will be different, because air is denser than steam. On the other hand, compressed air is easier and cheaper to produce. Whistles like these take a lot of air (or steam) to produce a really powerful, far carrying note.

If you do make one of these whistles, go easy on it the first time you try it: you can mouth blow a whistle of this size very easily ('tho you won't get a great volume of sound), and if you hook it up to a shop air line, for example, **the use of ear protection** when you turn it loose would reflect an acute perception of your own best interests. You'd also do well to warn others in the vicinity ahead of time.

Some points about whistles of this type:

The inlet hole for the steam (or air) supply needs to have a greater area than the area of the circular slit through which the steam escapes, to make the whistle work.

The width of the slit can be anything from say 0.005" wide to 1/16"; the wider the slit, the more air or steam the whistle will consume.

The normal practice in making whistles of this type is to make the ID of the bell equal to the OD of the annular slit in the bowl, so that the knife edge of the whistle bell or tone chamber is directly over the outer edge of the slit.

> (You will note that this would not be the case with the dimensions given for the whistle shown here. Just why the little Lunk did not follow this rule I don't know. If I were making one, I would be inclined to try it first the way it was supposed to be (i.e. with bell I.D. equal to O.D. of the annular slit in the bowl) and if it did not work well, I would then re-machine the bell to the dimensions given here. On the other hand, the dimensions given are those I measured, and that whistle did work.)

The bell must sit concentrically over the slit, hence must be accurately machined and nicely fitted to the stem of the whistle so that this is so, regardless of adjustments made as per next paragraph. In larger whistles, a "spider" is often fitted near the bottom of the bell, to keep it concentric with the slit.

Final tuning of the whistle is done by screwing the bell up or down along the stem until the tone is to your liking. The minimum distance from the bowl to the knife edge of the bell is 1/4 of the bell diameter.

I have spoken of the whistle bell as having "a knife edge". This does not mean that the bottom of the bell should be machined to a sharp cutting edge. If the actual edge is 1/16" wide, the whistle will still work.

If you care to research the matter of how to size a whistle to produce a particular note, you could then make what is known as a chime whistle: 2, 3, 4, or even 5 whistles, each producing a

A REPLICA LUNKENHEIMER STEAM WHISTLE

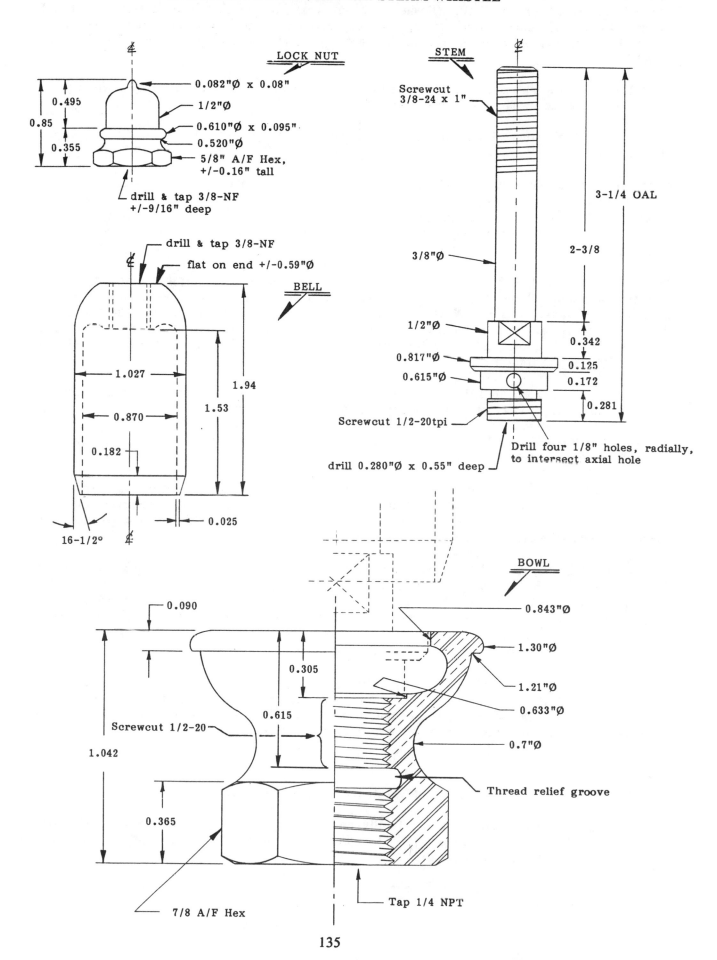

different note. If the several notes comprise a pleasing musical chord*, you could have a most unusual car "horn", or door bell. (I wanted to put some info on that idea herein, but when I asked a famous air horn manufacturing company here in Vancouver if my calculations for a series of 3 particular notes were correct, they got cagey and would only say I was "on the right track".)

    *(Try C, D#, & A#, a combination used on some locomotives of the US Army's Military Railway Service in North Africa during WWII.)

Now here's a little story about whistles - absolutely true so far as I know - that'll give you a chuckle:

Years ago, a fella we'll call "Benny" was driving down from a northern Vancouver Island town to catch a ferry over here to the Mainland. It was early on a Sunday, and the road was virtually deserted -- except for a police patrol car, which Benny overtook. For reasons best known to himself the cop prevented Benny from passing for a good many miles, speeding up where Benny could pass, and slowing unduly where passing was unsafe.

Benny restrained himself for perhaps 20 miles, until they came to a level railway crossing with 3 or 4 sets of tracks. When ye puttering policeman was right in the center of the crossing, Benny gave him a blast on his horn. Now Benny's horn was no ordinary horn - Benny had a full size 5-note locomotive air horn under the hood. And it worked, too. Thus when Benny "gave him a blast on his horn", I hope to tell ya the cop heard it, in living color. He panicked, and stalled the car right on the tracks.

Benny drove around the stalled vehicle and proceeded on his way. Sure enough, in a minute or so here comes the cop, waving his Christmas lights. Benny pulled over, and the cop pulled up behind him, but stayed in his car, so Benny got out and walked back to see what the cop wanted. Cop asked him why he blew his horn. Benny said he thought he'd seen a dog, and it just felt like the natural thing to do.

Cop said he wasn't supposed to have such a horn in his car, but Benny knew the law was silent on whether this was permitted or not, and he so informed the cop. He then told the cop what he thought of his driving habits, and that he was going to get back in his car and go about his business. Which he did, with no more bother from the cop.

A few days later the top RCMP brass for the Vancouver area came to visit Benny at his place of business, and they had a chat. The officer said that his man had admitted that he had, by his driving, been harassing Benny. He then told Benny that he should remove his horn from his car, or he was going to get ticketed by every RCMP officer who saw him make the slightest traffic infraction. Benny said, ok, he would remove the horn... "But tell me something: why wouldn't your man get out of his car, when he pulled me over?"

Seems the dawdling cop felt much like Bob Hope did during the filming of "The Road to Mandalay," when the bear walked into the railway car and stood on top of Bob, who had taken refuge under the carpet with someone equally famous. The bear said "woof", and Bob Hope said later, recalling this: "Well right there, we were lookin' for a laundry."

A CHRISTMAS GIFT SUGGESTION

If you're looking for a gift idea for family and friends, you might like the idea of giving something made in your own shop. If so, I have an idea which I think you might like, and that I am certain would delight any recipient.

This past summer I was given a wind chime - one of those things you hang outside, and it tinkles intermittently in the breeze.

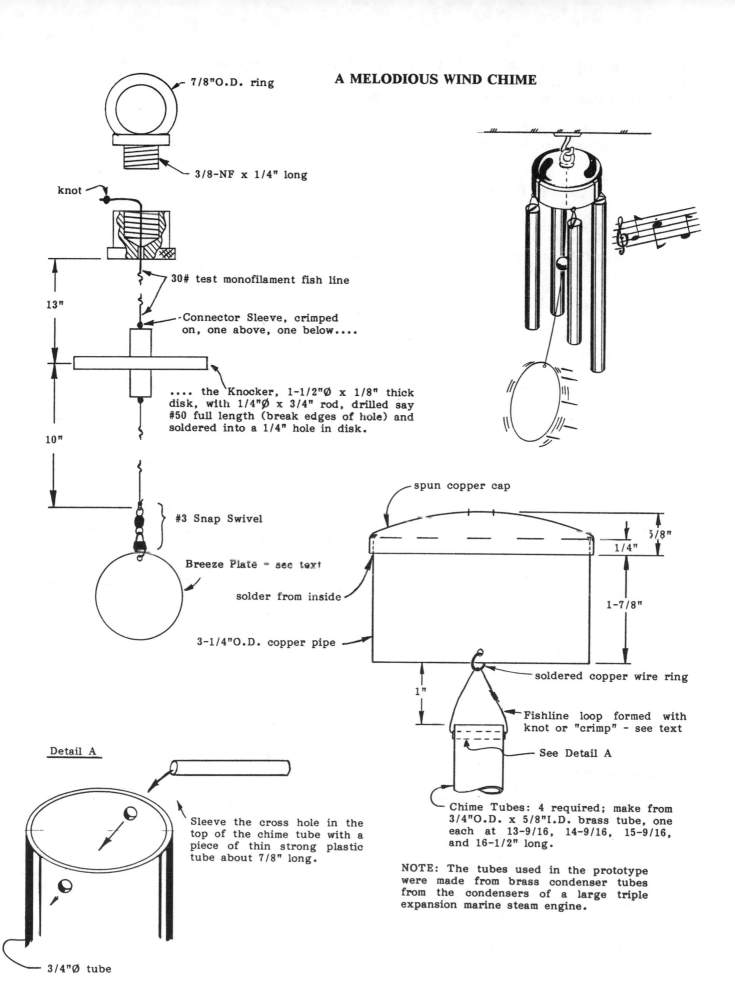

7/8"O.D. ring

3/8-NF x 1/4" long

knot

30# test monofilament fish line

13"

Connector Sleeve, crimped on, one above, one below....

.... the Knocker, 1-1/2"Ø x 1/8" thick disk, with 1/4"Ø x 3/4" rod, drilled say #50 full length (break edges of hole) and soldered into a 1/4" hole in disk.

10"

#3 Snap Swivel

Breeze Plate - see text

solder from inside

3-1/4"O.D. copper pipe

spun copper cap

5/8"

1/4"

1-7/8"

soldered copper wire ring

1"

Fishline loop formed with knot or "crimp" - see text

See Detail A

Detail A

Sleeve the cross hole in the top of the chime tube with a piece of thin strong plastic tube about 7/8" long.

3/4"Ø tube

Chime Tubes: 4 required; make from 3/4"O.D. x 5/8"I.D. brass tube, one each at 13-9/16, 14-9/16, 15-9/16, and 16-1/2" long.

NOTE: The tubes used in the prototype were made from brass condenser tubes from the condensers of a large triple expansion marine steam engine.

137

The difference between this one and others I've heard is that *this one has four of the most delightful and enchanting bell-like tones you could ever hope to hear*, whereas most wind chimes are about as musical as a bunch of tin can lids hung out to dry.

The chime tubes are hung from the suspension cap with fishline, per drawing, the ends of the fishline being fastened with something known as a "Crimp" or "Connector Sleeve", obtainable from fishing tackle outlets, as is the #3 Snap Swivel called out just above the breeze plate in one of the drawings. You'll need 6 crimps, one each for the 4 chime tubes, and two to position the knocker on the piece of fishline that carries the breeze plate.

You may have difficulty finding suitable material - 61 inches of **heavy walled brass tubing** - for the chimes; but look around for it - the resultant "music" is worth the hunt. I have heard that copper tubing (e.g. copper water pipe) can also be used, but whether it would sound as nice or not I don't know. I have seen chimes similar to mine, but having a wooden ball instead of a metal disk for the knocker; the wooden knocker produces somewhat more mellow tones. If a leather-wrapped wooden knocker is used, the tones are softer yet.

The plastic tube or sleeve noted in Detail A has a significant effect on the tones produced by the chime tubes - if you omit it, you will not get as nice a sound. The first place to look for suitable material for these sleeves, I think, would be the little straws attached to those small "drinkin' boxes" in which fruit juice is sold.

The other parts of the chime are simple enough. The "breeze plate" can be made from sheet metal (e.g. brass) or wood; it should be about as big as a man's spread hand, and can be any shape you please - a simple square, a hexagon, the initial letter of the family surname, a turned disk with a machine-divided pattern of drilled holes, an eagle in profile, or...

The drawing shows a piece of 3"O.D. copper pipe, for the suspension cap, and that seems a good choice, as it looks nice, particularly if the soldering of the spun copper cap is done nicely, and the finished cap well polished. This part could be simplified: in other wind chimes I have seen, it was reduced to little more than two crossed sticks.

You may have a tough time giving away your first wind chime of this design if your wife hears it, so consider making at least two.

WANT TO MAKE SOME EXTRA MONEY FOR YOUR SHOP?

I have a feeling that this wind chime would sell very well if placed in the right kind of store, possibly a florist shop or plant store - you'll have to see to that side of it yourself.

I think a fella could make a dozen of these chimes in a day, maybe more, and he could probably make about $25 above the cost of his material on each one - the store needs a piece of the pie too, so you can't have the whole selling price to yourself, unless you want to open a wind chime store, and I do not advocate that!

Thus, if one could sell the output, it would make a pretty nice little basement sideline.

ANOTHER IDEA FOR MAKING MONEY IN YOUR SHOP

Here's another idea for the guy who wants something that he can make in his shop and sell:

How about fishing reels? Some years ago I considered this idea, and decided against it - after all, if you can buy a factory-made reel for $40 to $70, who would pay $150 for a hand made one?

But I know of one chap here in Vancouver who makes fishing reels, and apparently he has a two year waiting list of customers. The crux of the matter seems to be that he does not make them full time, hence the demand may appear greater than if he were working at this one thing full time. If he did, he might not be able to sell enough reels to make a living out of it. However, that might not be the case, either.... read on.....

I recently attended a gun show here, and talked to an exhibitor who was selling some nice looking fly fishing reels. He told me a chap in England turns them out at the rate of about 5 per day. Nice simple design, very nice workmanship, in bright or black anodized aluminum (your choice) and brass, with blued steel screws. The asking price was about $150 US.

Now obviously the maker won't be getting all of that, but let's say he's getting $100 of it. Or, if he does get the whole $150, that his costs eat up $25. I think he'd be doing well to turn out 5 of these reels every day, so let's say he makes 4/day on the average. (I may be wrong on that point - depends on the fella's equipment and the degree of jigging and fixturing he has made up, and other things.)

What would you regard as a decent annual wage for a guy who works for himself, loses no time commuting, has no boss, and does what he likes to do?

If I could sell my output*, and liked doing that sort of thing, I could put my nose to the grindstone for about 4 months solid, pay for a shop full of tools, and take the rest of the year off.

> *Selling the output, of course, is the key point: making them is a mere detail, by comparison. If you can't sell your output, there's no money in it, no matter how many you make or how fast.

How to start? Look at the commercial product. If you can do nicer work, and/or have an idea for a better reel than those you see, make a prototype, and don't waste a lot a time making it fancy, for it is a certainty you will want to make some changes after it is done. Make a second one, and make it good. Test it, or give it to a fisherman friend for testing. Make a third one, and make it better yet. Make some more - this is called "payin' your dues".

When you know you have a winning design, make a run of a dozen or so, and get them out where they are most likely to sell - maybe at the local sporting goods store. From there, go after the bigger stores that cater to the well heeled sportsman on holiday. This poor peasant passed through Lake Placid, NY one time, and poked his face into a store of just the right sort. The sales lady had a diamond in her ring about the size of a golf ball, and when I asked the price of something - well! I might as well have asked her if she had fleas, from the look she gave me! Obviously it was one of those places where, if you have to ask the price, you shouldn't have come in.

> EXTRA! At pages 144, 145, 160 and 165 of the January 1951 issue of *Mechanix Illustrated*, you will find complete working drawings for a nice looking fly reel. Thanks to Tim Smith for sending along this reference.

Late addition to the above: About a month after writing the above, I poked my nose into a fishing tackle store in North Vancouver. The storekeeper asked me if there was anything I wanted. I said I was a machinist, and asked him if he thought there might be much of a market for handmade fishing reels.

Well! - the guy treated me like visiting royalty! He showed me a reel of a type much in demand, but the price of which is pushing Cdn$350 (and will soon be higher, with the exchange rate going the way it has since done.) Could I make such a reel? Yes, I could. (It was a Hardy "Silex". Other fancier Hardy's would command more money; more work to make, too.)

"And a lot of Hardy reels break just here: could you repair such a break?" Yes, I could. "And could you make replacement spools for such reels?" Yes, I could do that too - how many would you want - 20? Answer: "This one store couldn't take that many, but 20 could be spread between this store and several others if the salesman who calls on all such stores in this area were to spread the word that replacement spools were available."

He then showed me a reel made by a machinist in England - it was the same maker's work I'd seen at the gun show previously.

I myself am not interested in pursuing this type of work, but the point that struck me was that there was a bagful of work here if a fella wanted it. (If interested in fishing reel & rod repair, see pages 396-408 and 443-445 of the book *Gunsmith Kinks* (Vol.1), from Brownells.)

A BRASS KALEIDOSCOPE

In September '87 my wife and I drove down to Port Townsend, Washington, where, in a gift shop, we saw several kaleidoscopes -beautifully made, lovely to look at (and through) and wondrously priced - $85 and up.

A few months later, I happened to come upon the August '87 issue of a little magazine called *Woodsmith*; in it was an article on how to make a kaleidoscope. From a quick perusal of the article, making one was obviously a simple matter, and the timing was right - I would make one for Margaret for Christmas! Naturally, I would use metal instead of wood.

In the plumbing supplies section of a local hardware store, I found a perfect source of thin walled brass tubing in the form of an item called a "tail piece." It looked pretty nice after a cleanup with steel wool followed by Brasso. I cut the flared end off with a fine-toothed hacksaw and filed the cut end square. (It could have been machined square in the lathe, but holding the thin walled material would have presented problems. (Not insurmountable problems, just problems, such that hand filing was easier; see below.)

The scrap box yielded a piece of aluminum big enough for the eyepiece and the two rings which hold the two glass disks in the opposite end.

The width of mirrors required was found by means of a large scale drawing. Three strips of mirror glass (1" wide, 7.4" long, 0.116" thick) and two glass disks, one frosted on one side, were procured from a local glass supply outfit.

Note: you may find that your glass supplier will want a template to work from in making the glass disks. The shop I dealt with asked for a thin cardboard disk the size of the disks wanted. I made two sizes of disks, and told them that the glass disks should be somewhere between these two sizes. So they cut me *four* disks - two at one size, two at the other. Now I need 3 more strips of mirror and another brass tube!

The rest was fairly easy - I just turned up the aforementioned eyepiece and two rings, with the application of some more Brasso to give the visible aluminum parts a high polish.

A few pieces of junk jewellery were broken up and loaded into the business end of the K'scope, and a couple of sighting shots were taken through it. I hope to tell you, my k'scope was every bit as posh, outside and in, as the $85 ones. Come Christmas morning, Margaret was delighted.

"Well, what's the point of all this?" you may ask. Two things:

1. If you want to make an unusual gift, a k'scope is hard to beat: it's an attractive conversation piece which almost anybody would be pleased to receive as a gift.

2. I think that if a fella were so inclined, he could make some serious money making k'scopes.

The unit cost would go way down if several were made at one time - certainly the strips of mirror and the glass disks would cost less if cut in larger numbers, and copper or other tubing could be bought in longer pieces, to be cut to length at home.

If interested in k'scopes as a money making proposition, make up a prototype, and take it around to several gift stores or similar outlets; you'll soon get a pretty good idea of the likelihood of selling a bunch of them.

Some stores will want to operate on a consignment basis, paying you if and when they sell them. In this case, you should figure to get maybe 70% of the selling price. If the store is willing to buy outright from you, you will get less - probably more like 50% of the selling price. But this is probably the better way to operate if you can: you get immediate payment, plus the store then has an incentive to sell them. Above all, don't let yourself get caught with a bunch of un-paid-for k'scopes in the hands of somebody who goes bankrupt - chances are you will never recover them or the money owed you for them.

I figure a guy could crank out k'scopes in about 20 minutes each, given a few simple fixtures to hold suitably chosen materials. 100 k'scopes at $50 each is $5000 - not bad for a week's work, or even for 2 weeks' work, if it should happen to take you twice as long as I figure it would. Made in numbers like this, your cost per scope should come down into the $3-4 range, I think, so you should come out of such a project pretty handsomely.

As with fishing reels or anything else, the main trick is to sell your output. I would be inclined to make one, then a run of maybe a dozen or so, and see how they went. If they sold well, I'd tool up and make a hundred, if that was in line with the numbers my retail outlet(s) wanted. Timing? Get 'em ready to deliver in time for the tourist season, or about mid-October, when things are hotting up for Christmas.

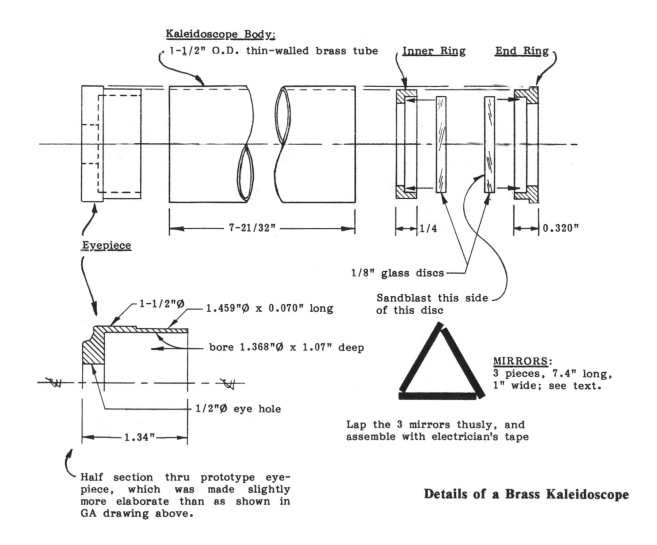

Details of a Brass Kaleidoscope

141

Some construction tips

You will need to know the thickness of the glass from which your mirrors will be cut, as this and the Eyepiece ID (1.368" on my drawing) will dictate the width of the mirror strips. If your brass tube ID is other than 1.458", (such that your Eyepiece ID will not be 1.368") and/or your glass is not 0.116" thick, draw a circle about 5x the diameter of your Eyepiece ID, measure it, and figure out your drawing scale. Then, knowing the thickness of your supplier's mirror glass, cut three strips of paper to scale thickness, with ends trimmed square, and all the same length (= width of the mirrors).

Overlay these on the circle drawing in a "lapped" triangle, as shown, and trim to such a length that the resultant fits comfortably in the circle, not forgetting that there needs to be some clearance between the mirror tube and the Eyepiece ID, and that this clearance, (and between the mirror tube and the brass tube) is readily taken up with cardboard wrapped around the mirror tube on assembly, as noted two paragraphs below. When satisfied, measure the paper strips, and divide by your drawing scale to find out what width to have your mirror strips cut.

To make the triangular mirror tube, lay one strip of mirror on the table across two pieces of tape, position the second mirror strip approximately, then raise the tape to stick to it. Add the 3rd mirror as above, and tape, then run a piece of tape full length of each corner of the triangular tube.

The mirrors, taped thus into a triangular "tube", will not be a snug fit in the brass tube. Wrap a 1/2" wide strip of paper and/or thin cardboard around the mirror tube at each end to take up the clearance between the mirror tube and the two interior diameters of the K'scope.

Use busted up junk jewellery (rummage sales, etc.), plus the odd colored button, etc. as "ammunition" for your k'scope. It is preferable to have most of the jewels transparent, rather than painted on one side, and naturally there should be a variety of shapes and colors.

The finished kaleidoscope, shown here on top of the Toolmaker's Block described at about page 20. At right is a photo of a typical k'scope pattern. Obviously, much is lost in the translation from living color to black & white.

Retain the glass disks in the Inner Ring and the End Ring with wax or glue. The Inner Ring should be pushed into the K'scope body tube about 1/2"; then the junk jewellery, then the End Ring with its frosted glass disk. If for "home consumption", leave the K'scope so it can be disassembled for cleaning of the mirror tube. Alternatively, a piece of glass in the eyepiece end would be a good idea, to exclude dust. If making for sale, glue the End Ring in solid.

Add any fancy work you like - knurling, etching, exterior bands of leather, ribbon, wire, sheet copper, etc. Polish all parts, lacquer the brass tube (or be prepared to re-polish it from time to time), assemble, and make a little display stand (not shown).

Cutting the tube for the K'scope body

The thin walled brass tube for the k'scope body would be difficult to hold for machining the ends square without special tooling. If you are making only one or a few K'scopes, use a very fine toothed saw to cut the brass tube; use the end of a purpose built miter box as a guide for the saw, and subsequently for the file with which you finish the job. Hold the brass tube in place in the miter box with a little wedge and a board.

If making a bunch of K'scopes, it'd pay to make or buy a mandrel onto which a piece of tube, roughly cut to length, could be slipped for finish machining of both ends.

If making a mandrel for this purpose, minimize the amount of precise machining required by making the mandrel undersize over most of its length, leaving a dead-to-size portion at each end, just inboard of where your clean-up cuts will be made on the workpiece. Then think about "rubberdraulic" lock-up of the workpiece: axial activating screws at each end operating on a silicone rubber filling in axial holes connecting with cross holes in the dead-to-size portions of the mandrel.

A JAR OPENER FOR THE LADY IN YOUR LIFE

Tim Smith sent me drawings of a simple jar opener he made from the remains of a broken bandsaw blade. Actually he ended up making several, because first thing he knew, his girl friend wanted one like it, then her mom, and her several sisters, and so on. So while it is obviously practical, don't say I never warned you: this is a self perpetuating little sucker, unless you run off half a dozen at one go.

The Smooth Jaw, which is not detailed separately on the accompanying drawings, is made from 3/8 x 3/4 x 5" CRS or aluminum. The use of slots instead of holes for the three 10-24 x 5/8" button head cap screws gives a certain amount of adjustability, which may be desirable. By the same token, it might be a good idea to locate the Toothed Jaw just a little farther over to the left, to increase the capacity - I found a few jar tops around my place that would have been too big for this unit to accommodate, if made exactly as drawn.

As can be see from the sketch at lower right of the drawings, the finished Jar Opener is mounted to the underside of a wall-mounted kitchen cabinet, with the wide end toward the operator.

NOTE: As this project involves drilling 3 holes through a piece of bandsaw blade, this looks like about the right place to fit in a tip from Brian King, a Californian who passed along several good ideas - the others appear elsewhere.

TIP: To spot anneal a band saw blade, cut about 1-1/4" off the sharp end of a 3-1/2" common nail, chuck in a drill press with the point down, and use as a friction spot annealer! I tried this, and believe me, it works.

Clamp the saw blade down solid, of course, and put a piece of wood underneath it so the heat does not get drawn off by the drill press table. Turn on the machine, and bring the nail down into contact with the saw blade. Increase the pressure till the tip of the nail goes red hot and then melts - the molten material will likely weld itself to the saw blade. When you look under the blade, there'll also be a charred spot on the wood. And when you turn the blade over, you will find a discolored spot on the underside - that blade is annealed! Centerpunch on that little spot, and the c'punch will dig right in, which it likely won't elsewhere on the same blade! After this spot annealing (or should that be spot *annailing*?) you can use a hand drill to put a regular twist drill thru the blade like it was plywood.

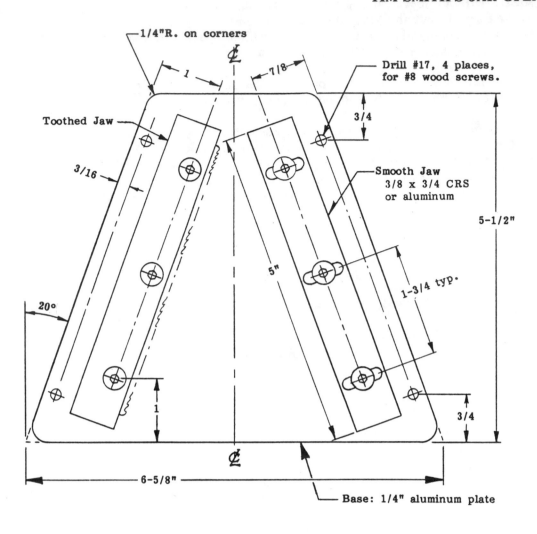

1/4"R. on corners

Drill #17, 4 places,
for #8 wood screws.

Toothed Jaw

Smooth Jaw
3/8 x 3/4 CRS
or aluminum

3/16

20°

Base: 1/4" aluminum plate

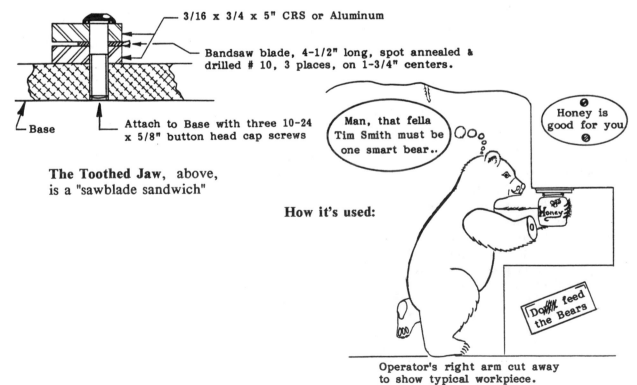

3/16 x 3/4 x 5" CRS or Aluminum

Bandsaw blade, 4-1/2" long, spot annealed &
drilled # 10, 3 places, on 1-3/4" centers.

Base

Attach to Base with three 10-24
x 5/8" button head cap screws

The Toothed Jaw, above,
is a "sawblade sandwich"

How it's used:

Man, that fella
Tim Smith must be
one smart bear..

Honey is
good for you

Honey

Don't feed
the Bears

Operator's right arm cut away
to show typical workpiece.

144

The World's Smallest lathe

Tim Smith sent me a photo of what appeared to be a 1905 Bardons & Oliver turret lathe, but it was sittin' beside a 22 cent US Postage stamp! A fella by the name of W.R. (Bill) Robertson, of Wheaton, MD made it. It is an almost exact scale model, right down to the back gears etc. He has also made a milling machine and some other machine tools, all on the same scale. Bill calls himself a miniaturist, as he accepts commissions to make miniatures*, mostly in wood, etc., of things like well known pieces of furniture, clocks (e.g. a 5" tall grandfather clock) and so on.

*Not stuff like the lathe - that he did for himself, for fun.

The above photo is from a color shot Tim Smith sent me, possibly one he took of a photo he saw somewhere; in any case it had some flaws in it (from the film processing, maybe), and it loses something in going from color to b&w here, but it's the best we got, so feast your eyeballs on...... **The World's Smallest Lathe.**

Most of us would not care to tackle a project like Bill's tiny lathe, but it does suggest an interesting avenue of pursuit. You may have seen an article in the Jan/Feb 1985 issue of *HSM*, at page 44, wherein Philip Duclos showed pictures of a 1/8 scale copy he made of his 5" milling machine vise. He does not say what the overall dimensions were, but the jaw width would be 5/8". Cute as a bug's ear!

If this idea appeals to you, take a look around your shop, and you may spot some tool or piece of equipment which would make an excellent subject for reproducing at some smaller scale. Someday I plan to make a copy of my 3.75"Ø Rotary Table at about 1/2 full size. It'd end up 1.875"Ø, with a 2" sq. base - cute, and what better for a paper weight for a machinist's desk?

Footnote: Since 1920, three model engineers have reported making an exact scale model of their lathe, and have shown photos of their projects in *Model Engineer*. The first was a round bed Drummond lathe, the second a flat bed Drummond, and the third a Myford ML7. Each of these lathes was, in turn, and in its day, the acme of the English model engineer's lathe, and interestingly, all were built to the same scale: 1/3rd full size; they would have been useful when done, as well as fun to make.

AN INGENIOUS HANGER FOR SHOP DRAWINGS

Tim also sent me drawings of a simple drawing hanger he made. It permits instant fumble-free hanging of shop drawings. I like it, and I think you will too. Make a couple of pairs. Attach two to a clear space on the shop wall, or to a 1/2 x 2" stick of wood, which suspend from a nail in the rafters. Mount the other pair where you do your drafting. True, a spring steel paper clip will do the same job, but such clips take one or both hands to operate, depending on how they are mounted. Besides, anybody can buy a spring steel paper clip - who but a machinist could make/own such a wondrous device as this? Your wife might like one as a recipe holder.

A GRAVITY OPERATED DRAWING HANGER

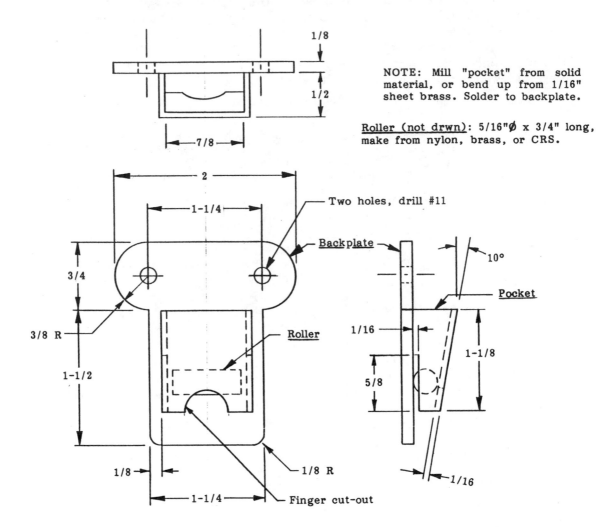

NOTE: Mill "pocket" from solid material, or bend up from 1/16" sheet brass. Solder to backplate.

Roller (not drwn): 5/16"⌀ x 3/4" long, make from nylon, brass, or CRS.

1/8

1/2

7/8

2

1-1/4

Two holes, drill #11

Backplate

10°

Pocket

3/4

1/16

1-1/8

3/8 R

5/8

Roller

1-1/2

1/8

1/8 R

1-1/4

Finger cut-out

1/16

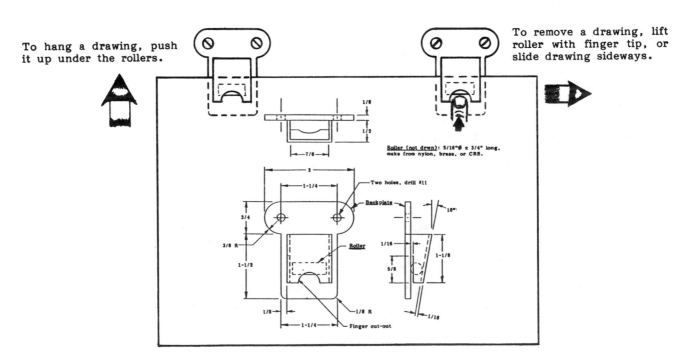

To hang a drawing, push it up under the rollers.

To remove a drawing, lift roller with finger tip, or slide drawing sideways.

IMPROVEMENTS TO THE DRILL SHARPENING JIG

I had a visit in March '87 from Ted Lusch of Mt. Vernon, WA. He happened to be the buyer of the very first copy of **TMBR** which was sold. He brought along his version of the Drill Sharpening Jig shown at page 26 of **TMBR**, and left it with me so that I could give it the once-over with my calipers.

Ted's modifications, shown in the drawing below, would alter the drill point angle produced, but this is not detrimental. Ted lays a piece of 400 grit wet or dry paper down on a hard, flat surface. On top of the abrasive paper he puts a piece of thin feeler gage stock. He then sets the DSJ down on top of the feeler gage stock so the tip of the drill to be sharpened contacts the abrasive paper. Now, the wheels roll on the abrasive paper, and the only other part of the Jig that is touching anything is the part in contact with the feeler gage. Thus the Jig is completely prevented from rubbing on the abrasive.

I like it, and it looks neat, too!

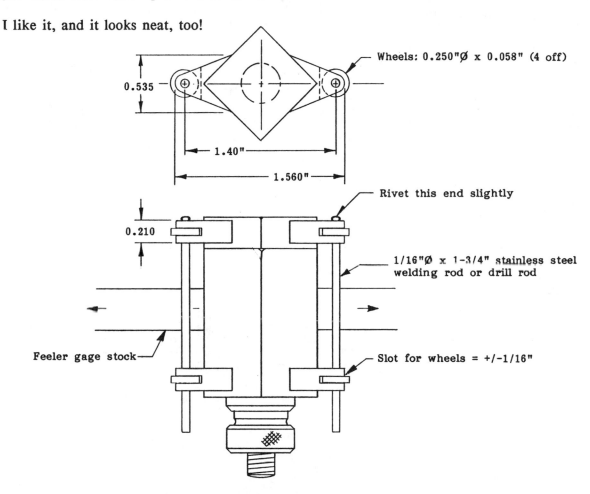

GUY'S COMPREHENSIVE INDEX TO THE BEDSIDE READERS

The best way to find what you're looking for in The Machinist's Bedside Readers is to look it up in "Guy's Index". 10+ pages. It's all alphabetical, goes by key words, and tells you which page to go to, in which book. Covers TMBR#1-3 and "Hey Tim...". Just ask for "Guy's Index" Price: US$7. *(in Canada: C$9.95)*

and............

GOODIES FOR CLOCKMAKERS!

We now offer a line of detailed working drawings for 7 different **clock movements**, a neat **sundial**, 15 **clockmakers' tools** (including a small clockmaker's lathe!) you can make, and more. **Phone, or ask for "the clock stuff" when you write for our regular catalog.**

THE HARALSON HOSE END

One day Bob Haralson showed me a bronze garden hose nozzle, and, as you might expect, there was a story behind it. Shortly after WWII a fella came to Bob with a drawing of a hose nozzle. Would he make a prototype? Bob did, and it worked as its designer had figured it would; thereafter he brought Bob pick-up truck loads of 1-1/4"Ø naval brass bar stock, from which Bob turned out hundreds of these nozzles; the designer sold them to gas stations all over the Western US in the late '40's and early '50's.

Why would this nozzle be of particular interest to a gas station owner? Well, you need a water pressure of about 45 psi minimum to make it really strutt its stuff, but if you have that or more, it'll cut oil and grease off a concrete floor like nuthin' you ever saw, usin' nuthin' but cold water outta your garden hose.

Bob loaned it to me, and I took it home & tried it, but I didn't have quite enough water pressure to make it get right down and boogie, so I took it down to the local fire hall. Well of course there's nothing so interesting to a fireman as a new nozzle, so the guys dug up an adaptor, and we hooked it to a 2-1/2" fire hose.

Unfortunately, there was no grease- or oil-on-concrete in sight to try it on, but I hope to tell you, that nozzle put out a jet of water that would be without equal for cleaning mud, etc. off the underside of a car - but you sure wouldn't want to play it on rust spots or loose

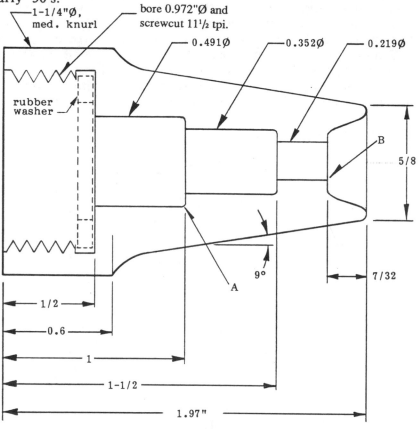

NOTE A: There was a slight radius at the bottom of each step in Bob's nozzle. However, the exact profile here is not critical to the nozzle's hydraulic characteristics. What **is** wanted is a sharp, square shoulder at the entrance to each step in the bore, as shown.

NOTE B: There should be a nice sharp edge to the exit orifice; beyond point "B" the profile of the muzzle of the nozzle is "in the wind", and therefore not critical.

paint! I very briefly put my hand into the jet about 2 feet from the nozzle, and it felt like getting hit by a bunch of needles.

The nozzle Bob loaned me was that first one he ever made. Some years ago - long after he'd ceased to manufacture them - he loaned it to a friend who wanted to hose mud, etc. out of the radiator fins of a big crawler tractor. Thereafter, for as long as Bob lived in that town, that guy would use nothing else.

You might want to make one. I think if a fella's water pressure was lower than 45 psi he could scale down the interior Ø's and get the same effect. If you try this, leave the outside as a straight cylinder during the experimental stage so you can rechuck it for further machining if necessary.

Another interesting possibility: put it on the business end of one of those hi-pressure washer units that are now common for cleaning concrete, stucco, etc. If you do, it would be prudent to throttle back the pressure initially, until you see what 50 or 100 psi will do, before trying it at 2500 psi! And I would be very careful what I pointed it at: even at 100 psi, I think it'd strip the shingles off a roof.

Happy squirting.

OUTSIDE READING

As in **TMBR**, I will provide some references to interesting/useful info in other sources. For those who don't like this, (there were 2 or 3 readers who criticized the presence of references rather than "hard copy" in **TMBR**) all I can say is that not everything can be done as I might like to do it, or as you might like it done.

Would you like to know how to make
SMALL, FINE-QUALITY ALUMINUM CASTINGS?

If you have an interest in making small, fine-quality aluminum castings, you should get a copy of the article "Small Precision Aluminum Alloy Sand Castings" by Gerald Smith. It appeared as a three-part serial in the magazine *Engineering In Miniature* (EIM), starting at page 130 in the October 1984 issue, and continuing on into the November and December issues. The author obviously knows what he is talking about. He tells how to make a simple form of molding flask that is compact and durable. He goes on to tell how he makes high quality metal patterns and core boxes, and describes a device for withdrawing patterns with near zero draft absolutely vertically from the sand. (One could duplicate this device quite easily, and the description given is adequate to enable one to do so.) He then deals with foundry equipment - crucibles, furnace & heat source (propane) - sand types, molding and pouring procedures, and finally what types of metal to use, and where to find them.

I would say that anyone armed with this article plus *Foundrywork for the Amateur* and *The Backyard Foundry*, by Terry Aspin, (see **TMBR**, p. 134) would have all the info he would need to turn out small, high class aluminum castings.

Getting this article through your local Library on Inter-Library Loan does not look very promising, as no North American library seems to have EIM in their collection. Therefore, if you want a copy of the article, I have set up the following:

Send US$5 cash directly to the publishers of *Engineering In Miniature* (see Appendix) and ask for a copy* of this article. (Do not send a $5 check. The cost of cashing it is so high there's almost nothing left of it; EIM said explicitly in their letter to me they would much prefer payment in cash.) They'll send you a reprint of the article, plus a copy of the then-current issue of *EIM*, plus a list of all their other publications. Leaving aside these promised extras completely, I am certain you will consider your $5 well spent when in due course you get the copy of the article - it is **excellent**.

*NOTE: Don't just write and say you want a copy of the article on castings. Be explicit: ask for "a copy of the article 'Small Precision Aluminum Alloy Sand Castings' by Gerald Smith, a 3-part serial starting at page 130 in the October 1984 issue of *EIM*, and continuing on with Part 2 in November '84, and Part 3 in December '84."

* see note re this at page 202

SHOP MADE SPECIALTY HAMMERS

For a good article on how to make your own special hammers, e.g. for auto body work, see *Popular Science*, Dec. 1945. pages 174-78.

HOW TO MAKE YOUR OWN DECALS

Decals provide a means of adding identifying markings or decorative touches to just about anything they will stick to, curved surfaces presenting little obstacle. If you want the same markings in several places, or on each of several items, duplicate decals can be more economical than hand painting it on each item, one at a time.

For full details on how to make your own decals, see an article by William T. Roubal, in the Nov. 78 issue of *Live Steam*. Roubal tells you the whole business: how to make the simple equipment needed, where to buy the few tools and supplies you'll need to buy, and so on. The article is well illustrated, and if the subject is of interest to you, do see it.

ELECTROSTATIC DISCHARGE MACHINING

Popular Science for March 1968 contains an article on how to make an EDM machine (if you make one like it, you may want to wire it so it's not electrically "hot"). **Better still,** see the serial beginning in the June/July '93 issue of *Strictly IC,* as noted at page 78 in **TMBR#3.**

An EDM machine can disintegrate a tap that has broken off in a hole, without damaging the workpiece. You can also sink cavities of almost any shape into steel. Say you wanted to make a die in which to form something of an artistic or practical nature - say an inspection cover for a model engine crankcase. You could make a male electrode of the shape desired, and use it to cut a female version of same into the die steel, and there's your die. (That's a one-sentence summary of a whole industrial craft practice, so don't feel bad if you don't grasp the entire range of possibilities presented by an EDM machine from this little precis.)

THE TESLA TURBINE

Have you ever heard of the Tesla Turbine? Not so many people have, as have heard of the Tesla Coil. Whether you have or not, if you are looking for a different, compact and powerful-for-its-size steam powered prime mover, you should look into the Tesla Turbine.

This engine was invented by one Nikola Tesla, about a hundred years ago. When tested, a prototype unit only 2' x 3' x 2' put out 200 HP on 125 psi steam @ 16,000 RPM. It is a relatively simple engine to build, as there are no reciprocating parts, and the "turbine" consists of a number of identical plain flat discs closely spaced along the main rotor shaft. Steam enters the turbine casing tangentially, and exits axially near the center. What makes it work is simply the drag of the steam on the discs. The efficiency is quite high.

Full working drawings for a working model can be found in two magazines:
1. *Live Steam*, Nov. 84, p. 32.
2. *Popular Mechanics,* Sept. 1965, p. 188-192 inclusive

Incidentally, working drawings & instructions for another interesting little steam engine of unusual design are to be found in the July 1965 issue of *Popular Mechanics* Magazine. Bill Reichart shows a photo of one he made to this design in *HSM* for March/April '87, at page 32. And, if you want to build a small two cylinder steam engine of nice design, with a real nice "bark" to its exhaust, check out the drawings for the Weaver Launch Engine from Baycom, Inc.

SPECIAL NOTE: If you are interested in making model internal combustion engines, you should subscribe to a new little magazine called *Strictly I.C.*- it's good. (Tell 'em I sent you.) See appendix for address.

A COPPER PIPE SOLDERING TRICK

If you've ever had to solder a joint in a horizontal run of copper water pipe, you have probably experienced the frustration of residual water in the pipe draining back to the joint, preventing you from getting it hot enough to make the solder flow and give a sound joint.

What to do? The following kink was given to me by my friend Wally Galpin.

Get everything ready - surfaces clean & fluxed, and all in readiness. Then stuff a wad of soft white bread - from the middle of the loaf, not the crust - into the pipe and push it in a little way from the joint area. Do the same thing the other way too if water is coming back to the joint from both directions. Now go ahead with your soldering. The water will not interfere, so long as you don't waste a lot of time about it.

When you turn the water back on, the white bread will turn to mush and flush out through the tap. At worst you might have to take the tap apart to let the bits of bread (toast?) through. My man, this could save your buns someday!!

ROCKY TAKES A HOLIDAY
Paul Niedzielski
Fair Oaks, California

I am an engineer by trade, not a machinist, so I have not spent that much time in shops, even though I have a very strong aptitude in this direction and it is my hobby. However, early in my career, I worked as a technician in the R & D Lab of a wire enamelling mill for a major corporation. This tale, which has several readily grasped morals, happened there.

It was toward the end of the year when the Lab Director got a burr under his saddle and decided that if anyone wanted to take a vacation he had to take all he had accrued at one time. We had one machinist, an old timer, and a very good man, who took care of all our machining and special fixture work. He'd been around a long time, and if memory serves me correctly, he had about five weeks vacation coming.

Rocky looked over the Lab Director's edict, and said he would fix his wagon. That afternoon he put in his vacation request for the rest of the year and for all of his next year's vacation, which was about six weeks, following New Years. We joked that he would be gone from Thanksgiving until Easter, which wasn't stretching the truth all that much. That left us in a bind - Rocky's was not a union job, and all the mill machinists were union, so we couldn't bring one of them in. We only needed one machinist, so none of the other lab personnel were qualified to do this work. However, strange as it may seem, most everyone who has ever had a shop course in high school seems to think he is a qualified machinist.

As I said, old Rocky was a good man. When a job was brought in, he would put it on his desk. When he was ready to do the job, he sat down, put his feet up on the desk and looked at the print, sketch or what have you. Then he would look out the window, then back to the print. When he felt he had studied the print long enough he would typically go to his toolbox, remove a few items, arrange them on his desk and return to studying the print.

Next, stock would be drawn from stores, and then he would spend a little more time studying the print. At this point he would usually scratch some notes on a small note pad. While all this was going on he would have a cigarette or two. More tools appeared on his desk. Finally he was ready.

The stock and the tools on the desk were allocated to the various machines, where they were put in order in a small tray kept there for that purpose. It seemed like nothing at all was going to happen. Then he went to work. As soon as he touched a machine you knew you were watching a master at work. Parts appeared under his hands like magic, in no time at all, to print, finished and looking pretty.

Using this approach Rocky got out a tremendous amount of work without ever appearing to move. He did, of course, but always with a plan. Now however, he was on vacation, seemingly forever, and we were between the proverbial rock and a hard place.

Al was an older engineering type who had taken a shop course or two in high school, and along with everything else he knew a little about, considered himself an expert on tooling and machining. In fact, he wasn't loathe to offer advice or criticism whenever he caught anyone working on some small job in the shop. Besides, as he was prone to say, "Machining is easy, anyone can do it, nothing to it". Anyway, with Rocky gone, Al took it upon himself to do all the machine work.

It was a study in contrast to watch him at work. Whereas Rocky never seemed to move, Al was a whirling dervish of activity. He zoomed around the shop, as he did the whole lab, ricocheting off walls, with a cloud of dust behind him, always at flank speed. After a couple of weeks of this frenetic activity however, it became painfully obvious that little finished work was coming out of the shop, and what *was* coming out wasn't much good. Thus, anyone with a job in the shop began to check on it frequently, and a parade of disgruntled project leaders began to develop. Of course this slowed things down even further, even though Al increased his pace, now moving at a blur.

One day I showed up to make some small modification on my own. Al was resurfacing a soft rubber strip coating roll on the lathe. The drill was to put the roll - which, if memory serves me correctly, was about 12"Ø x 18" long with a shaft through it - into a freezer for a while, or dip it in liquid nitrogen to harden the rubber. It was then put between centers on the lathe and ground, using a toolpost grinder and fine power feed to remove the rubber. When the rubber softened, it was refrozen and the work proceeded until finished.

As I was going about my business, Al whizzed over to the lathe to check on the progress of his rubber grinding operation. One of Al's idiosyncrasies was that he liked to project a good image, so he wore a shirt and tie. He had a shop coat on, but somehow in his frantic haste, his tie ended up outside the coat. Now everyone knows that you don't wear loose clothes around moving machinery. This is Lesson One, Day One, in every shop course; and Chapter One in all the books. It is also the prudent and common sense thing to do. It is also true that if your life depended on catching your clothing on some moving part it wouldn't happen in a million years.

This day, however, Murphy and his infamous law stalked poor Al. I heard a strangled gurgle, turned around and saw Al's tie wrapping around the shaft of the rubber roll. I ran over to the lathe, which was a 14" Clausing with a panic bar, and managed to get it stopped just as his Adam's apple was ready to hit the shaft. Needless to say, our "Expert" took a lot of ribbing about what could have been a very serious incident.

As is typical of many people who like to dish it out, Al couldn't take the ribbing very well, so he responded with even more than his usual amount of bluff and bluster. A few days later, he had irritated me past endurance, so another fellow and I grabbed him and stuffed him head first in an empty fiber barrel, from which he had great difficulty extricating himself. Three days later he suffered a heart attack. Whether the episode in the barrel had anything to do with it or not one can never know, but I have felt guilty about my part in it to this day.

With Al in the hospital, things were now really bad. Several of us now had to try to fill in doing whatever we could to keep the backlog down, even though none of us were really qualified to do so. Out of this period came my lifelong interest in machining, an interest which has aided my career development as well as being my hobby.

Rocky eventually returned from his vacation and things settled back into a normal routine. Within a week of his return however, we received another edict from the Lab Director stating that in future all vacations were to be taken in short blocks; a day at a time was fine, long vacations now being subject to executive review.

The morals of this true little tale are obvious:

1. Plan ahead.
2. Motion ain't necessarily progress.
3. Safety first, last, and always; particularly over appearances.
4. Horseplay has absolutely no place in any work environment.
5. Last, but not least, NEVER underestimate the power of a wily old machinist.

"We lost that one sheave completely..........."
Paul Niedzielski

Another interesting incident happened while I was working at the same place, although I wasn't directly involved in it.

Many people are aware that finely divided powders, such as flour, in combination with oxygen in the air, can be explosive under certain conditions, even when they are not usually thought of as being combustible. Few people are aware that finely divided metals can also be combustible under the right conditions, even though they may know about the Thermite process for heavy welding. I have had 12 micron diameter Titanium (Ti-6A1-4V) fibers spontaneously ignite and burn with a brilliant blue-white flame. Fortunately, no one was injured, and fortunately for me, my boss was standing talking to me at the time so I did not have a credibility problem. However, that incident greatly impressed upon my mind the need to keep my shop mess cleaned up; even though I don't always strictly adhere to it. This incident is similar.

We had a vertical enamelling oven that was being converted from large rectangular wire to round. Enamel was picked up onto the wire as it entered the oven through a die at the bottom of the oven. The coated wire was then pulled up through the oven where it was cured. The cured but flexible enameled wire was then run over a big sheave and down the outside of the oven to the bottom, where it went round another sheave, reversing its direction, and went back through the enamel and on up through the oven again.

Many passes were required to reach the desired coating thickness. The sheaves were like big V-belt pulleys, about 48" diameter, and made of magnesium for lightness. The configuration of the groove on the pulley controlled wire location as well as enamel abrasion and was somewhat critical.

We had no facilities for turning a workpiece of this large a diameter, so the job was sent out. There was an "old time" shop close to us that had some big old equipment which could handle work of this size. It was a dirt floor place in a big old barn of a building, and was run by a crusty old fellow. Whatever other virtues this fellow had, housekeeping wasn't one of them. The lathe beds were almost obscured by the debris of a hundred previous jobs.

The owner of the shop proceeded to chuck and indicate one of the sheaves. When he was ready to start making chips, the engineer from our shop who was overseeing the job asked him if he didn't think he ought to clean off the lathe bed and surrounding area since magnesium was know to be capable of being pyrophoric. This set the old guy off, and our man was told it was his shop and he'd do what he damned well wanted in it. Besides, he'd been machining since before that "kid" was born and he'd machined lots of magnesium with no problem. If he didn't like it, he could take his job elsewhere. Period.

Our man shut up and the job proceeded.

Several sheaves were machined while the old guy mumbled under his breath. And then something happened. Whether the owner attempted too heavy a cut and got a red hot chip in the wrong place, or whether he unfortunately got the wrong mix of metals in his bed of swarf, or whether his luck just ran out is uncertain; for lo it came to pass that the mess of chips caught fire - and cut his lathe bed in two!

We lost that one sheave completely, and had to take the rest of the job elsewhere. Since that time I have had a healthy respect for seemingly innocuous little pieces of metal, and always try to clean up after each job I do.

ALICE LOSES HER SHIRT
Paul Niedzielski

Another story, not strictly about a shop, but which is absolutely true, as I was there to witness it, illustrates nicely a basic safety tenet.

I was working as staff engineer in a textile plant. One of the picker operators was a magnificently endowed - and bra-less - young woman. One day while operating her machine, she got too close and caught her shirt - which because of her figure hung out from her front anyway - in one of the rolls. She braced herself on the roll guard and wasn't hurt, but her shirt was torn completely off. She looked at the machine for a second or two, and then turned and walked, in all her majesty, to the personnel and safety office at the plant entrance. This took all of about 15 seconds. In that short space of time, we had 4 reportable injuries: two forklift accidents, one black eye from running into a support pole, and one fellow put his hand through the glass door of a recording pyrometer. Incidentally, all of those hurt were male.

This episode illustrates another valuable point: when you are working with machinery, always pay attention to what you are doing. Distractions, regardless of how distracting, can be hazardous.

Al ZUEFF MAKES A PROP SHAFT

I've mentioned Al Zueff elsewhere in this book. Al, who probably knows everything there is to know about being a tug boat captain, admits he yet has a few things to learn about the machinist's trade, but the following tale illustrates that he is a quick learner:

A fella came to Al with a messed up prop shaft, and a piece of material, and asked if he would make him a new prop shaft - the local machine shop wanted several hundred dollars to do the job.

"Sure, no problem," says Al.

A price was agreed, and Al discounted the importance of the fact that the material supplied - some stainless steel shafting - was 1/8" over the size of shaft the guy needed - he would just take a cut over it and bring it down to size.

As soon as the guy left, Al got on the phone to his machinist friend. "Hey! I got a prop shaft to make. It should be a piece of cake. Come on over and show me how it's done, and you can take 3/4 of the price." So his friend comes over, and they chuck the piece of material in Al's lathe and take a trial cut.

You guessed it - that was the toughest kind of stainless steel you could put a tool to! As Al says, "..the tool just bounced over it." Well, now what to do? Al's friend said to leave the job sit for the day and he would see if he could get a suitable piece of shafting elsewhere.

Next evening he shows up on Al's doorstep with a piece of polished and ground stainless steel shafting just the right size for the job. Naturally, it didn't take too long to turn a taper on the end, screwcut a thread, and carry out the various other ministrations required to transform the material into a very nice looking prop shaft for the customer's boat.

Next day Al calls the boat owner to tell him his prop shaft is done. The guy comes over and gets all starry-eyed over the beautiful finish Al has achieved in turning the material down to size. "That's real fine! Did you have any trouble getting it to come out so nice?"

"Oh, no sweat at all," says Al.

"Gee, it's beautiful! How did you ever do such a nice job?"

Al shrugs and allows as how it's easy when you know how.

"Well, I really appreciate this," says the guy. "Say! Would you like to have some more of that material? I could get you some more of it pretty easy."

Al told the fella he was generally pretty well fixed for material of that sort, so he'd pass on the offer, handsome though it was.

Like I said, Al is a fast learner.

"There's a Bridgeport in my Basement!....
and I'll tell you how it got there."
Tim Smith, Toledo, Ohio

Having put up with the frustrations of a small mill-drill for a number of years, I finally decided that the time had come to buy a full-sized Bridgeport-type vertical mill.

After much shopping around for the correct model, price, etc. I bit the bullet and ordered what seemed to me to be the best machine at the best price. I was on the way home when it suddenly sank in: I had just ordered something that was going to arrive on my doorstep on a truck in a crate with an all-up weight of 2800 pounds.

I had bought myself two problems:

First, how to unload this monstrosity from the truck?

Second, how to get it into my basement shop?

I had initially planned to install the mill in my back yard garage, but in my heart of hearts I knew that shuttling tools, etc., back and forth between basement and garage forever after would not be satisfactory.

No one I knew had any experience at moving a 2200 lb.* item down a flight of stairs into a basement, so I would have to figure out for myself what to do. The more I thought about it, the more I figured I better get help from somebody who knew what he was doing, and who had good insurance!

*The weight of the milling machine alone is 2200 lbs; crated weight is 2800 lbs.

Sure enough, under "Machinery Movers" in the Yellow Pages, I found a number of outfits listed. I struck paydirt on the first call - "Billy the Rigger" said he had made his living for the last 35 years doing just this sort of thing. He said he could move anything anywhere, and proceeded to tell me about a mill he had put into someone else's basement, but where the only access had been by an impassable spiral staircase.

"The guy sent his wife off on a shopping spree. We picked up the mill with a little rubber tired mobile crane, drove into the front yard on planks, rolled up the living room carpet, cut a hole in the floor, poked the mill in through the front door, and lowered it into the shop. The guy patched up the hole in the floor, and rolled out the carpet again. So far as I know, he never did tell his wife what we did."

Billy told me to have the mill delivered to his shop, where he would unload it, dismantle it to his liking, and then have his crew deliver it and put it in my basement.

A couple of weeks later Billy phoned to tell me the mill had arrived and was now dismantled and ready for delivery: when would be convenient to deliver it? We settled on 3 PM the following Monday.

Monday, sharp at 3, Billy's truck arrived, complete with power tailgate, three men, and my milling machine. The mill had indeed been taken apart - the base casting was sitting in one corner of the truck, and the rest of the machine had been broken down into separate pieces: knee, table, head, motor, etc.

The first order of business was to reinforce the basement stairs with 4x4's, of which they had brought along a good supply.

That done, the crew decided the big base casting should go in first.

Eyeing the heavy, awkwardly shaped base, I wondered to myself just how they would go about this little feat of magic. I was about to find out.

Two of the men pushed the casting onto an overgrown version of the sort of hand truck one might use to move a refrigerator. This one, however, was considerably heavier.

With two men on the back handles and one guiding the front, down the stairs came the base of my precious milling machine. Not fast, mind you, just slow and steady. The knee followed, then the head, motor and table in turn. Billy's men made it look easy!

The guys rigged a makeshift hoist, and with its aid the knee was re-attached, followed by the other parts. Wiring it up, levelling it, and dialing it in would be my responsibility, but there was my gorgeous new B'port-type mill, safely in place in my shop, with not a scratch or a mark on it.

I worked my toenails free from their grip on the insoles of my boots, and heaved a sigh of relief. The tension headache I had had for the past 3 days began to disappear.

I suppose most hsm's enjoy, as I do, solving problems, whether in connection with shop projects or otherwise, with the minimum of effort and expense. For me, half the fun is doing the maximum amount of work with the minimum number of headaches.

However, when it comes to installing a large, heavy, and expensive machine tool in a basement workshop, the "smart and easy" way to handle the matter, in my opinion, is to have a pro do the job. If I were faced with the same problem again, even having now seen how it's done, I would not attempt to do the job myself. The pro's have the equipment, the muscle, the know-how, ... and the insurance.

Although not exactly "cheap" in one sense, the realities of doing it yourself are these: if something gets damaged, it is purely your own bad luck; you also risk serious injury to yourself or friends (probably totally inexperienced in such matters) who've been conned into helping. "Sorry" can not restore a crushed hand.

It cost me just under $300 to have my mill installed in my basement, but when I saw it finally sitting exactly where I had known it should be, the cost seemed like a ham sandwich.

A MECHANICAL EL DORADO
- An afternoon's supreme enjoyment -
by Herbert Dyer
Reprinted, with permission, from the
August 15, 1935 issue of *Model Engineer*

A year or two ago I purchased a particularly fine lathe from an old gentleman living in my district. From his conversation I gathered that he was very interested in all mechanical pursuits, and had "many machines and tools". A glimpse of these marvels was either denied or else cleverly side-tracked, and so I came to suppose that perhaps that the old chap, now deceased, "had a wee bee in his bonnet".

Recently, however, I was visited by a relative of his who asked me if I'd care to look over the old man's kit and sort out the sheep from the goats, so to speak. I could also, in the capacity of friend and advisor, have first pick, if there was anything I fancied, before the dealers fell upon the spoils.

Well friends, I went along with this willingly enough, and what I saw you would be hard put to believe, had you been there yourself....

In a back room, spread upon the floor and on shelves, was as much gear - in new condition - as one would see in a modern tool dealer's store.... To enumerate a few items: a brand new Norvic hand shaper with swivelling machine vise; four or five other machine vises on the floor; three bench vises ranging from three to six-inch jaw width. Plus somewhere about a dozen lathe chucks from three inches up to seven - Cushman, Union and Crown, all new and in boxes. Die stocks, some with bronze stocks, stamped "J. Buck, London".* About half a hundredweight of new lathe tools, some in bundles of a dozen, and some loose. I know, 'cos I lifted 'em. Four drilling machines for hand bench use, a geared forge, a geared hand grinder and a power ditto.

*(Doubtless the writer's meaning here is a top quality item - perhaps even the very best make available at the time. GBL)

Turning about gingerly so's not to trip over anything, I saw a couple of 4" Round Bed Drummond lathes, new, but with items missing which doubtless could have been located had the time been available. Keat's V-angle plates, drill spindles, expanding mandrels up to about 2". Two or three small surface plates, and boxes of oddments which'd have taken hours to dig over - a delightfully interesting job.

There were hosts of drills in stands and not in stands, files, handles, and much else, all in glorious profusion. Incidentally I boned a set of high speed drills from 1/8" to 1/2" and a set of hand reamers up to 5/8", at my own price, which offer was very kindly accepted - really a silly price, but all that I could afford, having an eye for other items and not daring to blow further than my breath permitted.

... Before I'd had a chance to recover somewhat from this shock, I was transported in more senses than one to another building, inside of which the marvels of the first scene were as nothing. Surprise was heaped upon surprise literally - lathes, planers, wall mounted drill presses, surface plates, wood turning lathes. Everything was new, or nearly so. NEW!! ye mechanical salapeens! Think of it, and weep! Ali Baba's Caves or El Dorado wasn't in the picture 'long-side of this show.

Thick dust lay over everything. Fat, obscene-looking spiders gazed hostilely at me from their webs, which seemed to encompass everything in their silky, sticky folds. The very windows were darkened by the dusty accumulation of years. This somewhat awesome atmosphere was not lessened by the knowledge that I was in the interior of an old chapel, a fact I had not at first realized.

As I said before, scarcely anything had ever been used, and with one or two minor exceptions - cases of light surface rust only - everything was in new condition. Facing me as I entered was a fine little 9" swing Tangye lathe. Behind that again, a very large wood turning lathe. A little to the left front was a 7" swing back geared lathe - make unknown - very light but definitely in the precision class. A Firth hand planer occupied the left, alongside of yet another shaper on a barrel. On the right front, bolted to a pillar supporting the roof, was a blacksmith's geared drill press, and in the foreground a 24 x 48" iron surface plate, cluttered with castings and components of all sorts - box angle plates, face plates and a whole stack of grinding wheels up to 10". I bagged a box angle plate and two medium 8" wheels.

A new but incomplete Relmac lathe protruded from a side doorway, whilst in the same alcove were jammed a conglomeration of counter-shafts, treadles and bundles of MS bar.

The original cost of all this will never be known, I guess, collected and accumulated as it had been over a period of some thirty or forty years - as witness the Tangye lathe and Buck's bronze die stocks with blued and polished dies and adjusting screws, all fitted into solid mahogany cases.

From the venerable precincts of "ye olde chapel" I was guided through yet another doorway into an outside building where, on the floor in a state of ghastly neglect and rust, was all that remained of a 6" Barnes lathe which had only that morning been discovered, when it fell through the bottom of its rotten crate where it had lain for possibly 20 years. Mayhap it had tried to emulate the time honored experiment of "turning over in its coffin" ... Anyhow, there it was, slide rest rusted solid to the bed, headstock gears immovably rusted together, whilst the changewheels - also under a solid cover of rust - stood in a corner under the legs, the four-speed countershaft on top of the lot, partially obscured in a shroud of spider webs and dust. Once a beautiful tool, it was now scrap.

Here, methinks, is ample food for thought. Many of us, both professional and amateur, are hard put to it to get a reasonable outfit together, yet here, in a locality far from the haunts of the ubiquitous machine tool, I had found a small fortune in tools, in some cases gradually disintegrating under the insidious action of those arch enemies of ferrous metals - RUST and NEGLECT.

....I have heard many yarns about collections of "objects d'art": paintings, sculpture, clocks, cups and saucers, and antique snuff boxes. I have even heard and read of people so ill advised as to have hoarded their money - these were, of course, not model engineers. And I have heard, though never believed, of a man who bought tools simply to look at.

Now I have seen with my own eyes that at least one of these people did exist. I knew him. What is written hereintofore is only a list which I mentally compiled - much else being hidden in the general confusion of the whole - but absolute fact. Here was a man who never used the tools he bought. He bought them, unpacked them (when he remembered), enjoyed them, and took some of them to pieces in the unmechanical surrounding of an old chapel. He gazed upon them.... and bought more.

He sold me my big lathe, happily before Old Man Rust caught it in his clutches. That sale must have been for him like drawing blood.

He died, thinking possibly of yet another future acquisition. I am not as a rule at all retrospective, but it is very strange that this is the second occasion upon which I have been asked to "look over" the tools and gear of a fellow enthusiast who has gone from this world.

One day we will all go that path, and I often wonder if the future users of our various outfits will do them justice.... I rarely use tools which have come to me in this way without giving a passing thought for their late owners - old friends who handled them even as I do, and who delighted in the "feel" of a good tool. To use their tools is to shake their hand in friendship.

Three footnotes from GBL:

1. Herbert Dyer, who wrote this piece, was a master sheet metal worker. He lived in Mousehole, a little town at the extreme southwest tip of England - what the Brits call "Land's End". The locals pronounce the name of the town "Muzz'-ole". Dyer wrote some very fine pieces in *M.E.* in the late 1920's and later; his articles were good stuff, beautifully illustrated by his own hand. At page 134 of **TMBR** I recommended his book *How to Work Sheet Metal*, and I would reiterate that suggestion here.

2. One day each of us will pass on, as did the chap whose "kit" Dyer looked over as above. None of us will take anything with us, not so much as a 6" rule or a mike. Just what to do with one's accumulated tools may be a puzzle to some. In **The Bullseye Mixture**, elsewhere herein, you may discern what I plan to do.

3. Within the B.C. Society of Model Engineers, the shop tools and equipment of a deceased member are commonly auctioned within the Club, if this suits the member's widow and/or family. We also observe a minute of silence at the monthly meeting following the death of a Club member. I always wonder what goes through the minds of others during this period of silence - my own thought invariably is that some day these men, or others like them, will stand thus for me, and all the tools and stuff that I have bought or made will be used by someone else. I hope that person will enjoy them as much as I have.

Solomon, the wisest man who ever lived, had something to say about all this some 3,000 years ago. His comments were much more profound than my mumblings. You can read what he had to say in the Bible, in the book of Ecclesiastes, Chapter 2.

OSBORNE'S MANEUVER
Adapted, with permission, from an article
by L.C. Melton, in *Home Shop Machinist*, July/August 1988.

I've poked through machine shop books that go back to the 1850's, and I'd never seen this one before. **Osborne's Maneuver** is a fast and easy method for centering a round workpiece under a milling machine spindle without the use of an indicator. So far as is known, Herb P. Osborne deserves credit for originating it.

Suppose you have to center a rotary table under your milling machine spindle. Aside from the fact that you're gonna be cranin' your neck to watch the idicator, because half the time it's facing away from you, the T-slots in the edge of the Rotary Table prevent an uninterrupted sweep of the indicator foot around the job.

(Alternatively, the task might be to set up over the center of something round that you have grabbed in the milling machine vise. If the workpiece O.D. has no slots, holes or bumps to obstruct the indicator, well, you just lucky, is all.)

But let's suppose the task is to locate a milling machine spindle directly over the center of a rotary table. (See drawing.) What to do is this:

Bolt the Rotary Table (RT) down on the milling machine table where you want it. Orient the RT top so one T-slot is at about 1 o'clock/7 o'clock, (12 noon being near the mill's column, 6 o'clock out at the front of the knee) to get the ends of the T-slots out of your way.

Measure the diameter of the RT accurately. Let's say it is 8.000"Ø.

Stick an edge finder in a collet or a keyless chuck in the machine's spindle nose. Turn the machine on, and pick up any point on the rim of the RT with the edge finder. (It makes no difference whether you start with the Table feed or the Cross-feed, but let's say you start with the Table feed.) ANY point on the OD is fine, but somewhere near 2 o'clock is good.

Having picked up a point on the rim of the RT - the point labeled (a) in the first drawing - raise the quill, and zero the mill's Table feedscrew dial. (Or, if you have a digital read out, zero the Table axis read out.)

Move in over the top of the RT a distance equal to the radius of the RT plus the radius of the edge finder tip (i.e. 4.000" plus 0.100", = 4.100"), and stop at point (b).

Now go to the Cross-feed axis, and repeat what you did on the Table axis. Pick up the edge at (c), raise the quill, zero the mill's Cross-feed dial, move in 4.100", and stop at point (d).

One "step" is one pass on each of the two axes.

Run **Osborne's Maneuver** 3 more times, each time using first the Table feed, then the Cross-feed, and each time using the last position on one axis while picking up the rim of the RT on the other.

Notice how rapidly the offset from machine spindle to RT center is reduced:

At (a), when we pick up the RT edge, the offset is 4.100".

At (c), after the second move, the offset is only 0.550,54".

One step in Osborne's Maneuver has reduced the offset from 4.1" to 0.55".

After a 2nd pass on each axis, we will be at (h), and the offset will be only 0.000,17". The 3rd pass cleans up any remaining error, while the 4th pass is purely precautionary.

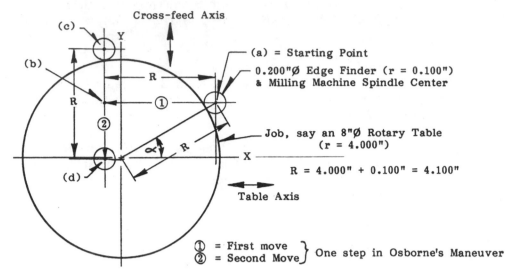

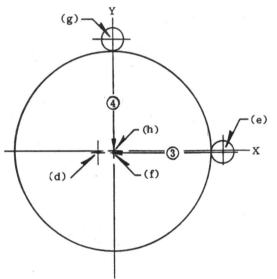

③ & ④ are a 2nd pass on each axis, and constitute the 2nd "step" in the 4 steps of Osborne's Maneuver.

Another example of where **Osborne's Maneuver** would be useful would be in setting up over a Toolmaker's Button which has been secured to a job and accurately located thereon.

Note too that the bigger the OD of the job being set up, the less accurate is this method. But don't let that bother you: I ran through the numbers for a workpiece having a 50" radius, (i.e. a very big workpiece) - after 2 passes on each axis (i.e. 2 steps), starting from about the 2 o'clock position on the rim, the spindle should be within 0.003" of the center of the job. The 3rd pass corrects this slight error. The 4th pass is, as noted above, precautionary.

If you want to investigate the math behind **Osborne's Maneuver**, make it easy on yourself by taking the radius of the edge finder as zero.

"Lautard's Maneuver"

I said above that when picking up the OD of the workpiece with the edge finder initially, "...any point on the OD is fine, but somewhere near 2 o'clock is good." Actually, assuming you are going to make your first move on the Table axis, the nearer to 3 o'clock you start - i.e. the smaller angle ∝ is - the fewer moves will be needed to center the work.

If you start with a point say 2 or 3° before the 3 o'clock position*, you will be centered after 3 moves: one move on the Table feed, one on the Cross feed, and one more on the Table feed (i.e. after only one and a half steps of Osborne's 4 step procedure). Run an extra pass on each axis as a precaution, if you want.

*To aid in setting the edge finder so close to the 3 o'clock position, use "Lautard's Maneuver": scribe a diameter line on the work (or use the end of a T-slot, if the job is a rotary table), and set this line approximately parallel with the milling machine's table axis by eye. Bring the edge finder to the end of the scribed line, then crank the cross feed handle out towards you maybe one full turn (i.e. about 0.2") for an 8"∅ RT, or less, if the work is of a smaller diameter. This gives you a very small offset from center one way at the start, hence, you get centered faster.

For those who have built themselves a
TINKER Tool & Cutter Grinding Jig

LUBRICATION AND GRINDING MACHINES
and some other tips

I had a letter from a guy who prefers to remain nameless, giving me some new and good ideas re the TINKER Tool & Cutter Grinding Jig. I will include them here in **TMBR#2**, for the benefit of all, because some of the info is useful not only around the TINKER.

1. Never oil anything around a grinder. The oil will hold grit, and make for greater wear. Instead, do this:

Clean all parts that move and then apply Molycote, GearMoly, or some other molybdenum disulfide product on all moving parts, then work them together for a minute or two. Take the unit apart, and wipe - not wash - all grease off. Then put the unit together again. The moly will work into the metal and act like a bunch of little ball bearings. This treatment will last maybe six months or more, at which time repeat the procedure. If the parts tend to rust because of the absence of oil, you may want to have them blued.

2. For the parts of the TINKER that require an engraved scale, apply instant (cold) blue to the engraved area, then rub with steel wool. This will leave blue in the markings, making them show up better.

3. Re Pivot Shaft (part #16): Put the End Cap (part #6) in its place in the Foot casting, screw Pivot Shaft in to full depth in the End Cap. Scribe a line where Pivot Shaft emerges from the right side of the Foot casting. Disassemble. Put the Pivot Shaft in your lathe and machine a shallow (but more visible) groove in place of the scribe mark. Now you can tell at a glance the amount of travel towards the grinder still available when setting up to do a cutter - better than bottoming out in the middle of a grind and having to reset the job.

4. Make Part #23 a little longer than in original drawings, or solder an extension to it. Drill and tap 10-32 to intersect the tapped hole for the Thumb Screw. With a leather or sheet metal punch, make a plug from a piece of sheet Teflon, nylon, leather, plastic bleach bottle, or like that. Put the plug in the hole, and follow with a set screw to push the plug against the Thumbscrew so it does not turn unless you want it to. When the plug wears out, make a new one. (Or: apply a little rubberdraulic thinking here too. GBL)

5. Two more places to buy #2MT collets are Travers Tool, and Industrial Pipe and Steel. (See Appendix for addresses.) These will be found to be draw-in type collets - i.e. they employ a draw-bolt from the rear - but this is no disadvantage.

"....I was so happy I could have cried....."

Sometimes fussy painstaking work, or just too much of all the same type of work, can "get to" a fella. This point was raised in conversation with Philip Lebow, who hangs his hat in southern California. I got a chuckle out of the following, which he sent:

"In one shop where I worked, I once spent 6 weeks full time doing the rough scraping on a big Landis thread grinder we were re-building. An old boy with magic in his hands did the final frosting and flaking. He was good. When the job was done, my hands were so sore I couldn't hold a mike for a week. The boss put me on a big Fosdick radial drill, rough boring big die sets. I was so happy I could have cried. But I did end up with some neat calluses that a scraper would snap right into."

161

A Handy Decimal Equivalent Chart

I saw a very good decimal equivalent chart in the July/Aug. '83 issue of *HSM*. It had been submitted by one James Hamill. Use it once to answer the question Hamill asked at the end of his few words which accompanied it, and you will fall in love with it. His question: **"Quick: How many 64ths is a bit smaller than 13/16ths?"** I have changed it somewhat, putting the decimal equivalents in bold type, and inserting commas after the 3rd decimal in them. Mr. Hamill, when asked permission to use this chart herein, asked that he be credited only with submitting it to *HSM*, and with gratitude to some unknown author.

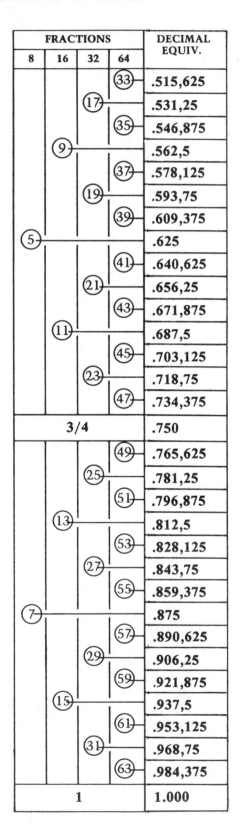

FRACTIONS				DECIMAL EQUIV.
8	16	32	64	
			1	**.015,625**
		1		**.031,25**
			3	**.046,875**
	1			**.062,5**
			5	**.078,125**
		3		**.093,75**
			7	**.109,375**
1				**.125**
			9	**.140,625**
		5		**.156,25**
			11	**.171,875**
	3			**.187,5**
			13	**.203,125**
		7		**.218,75**
			15	**.234,375**
1/4				**.250**
			17	**.265,625**
		9		**.281,25**
			19	**.296,875**
	5			**.312,5**
			21	**.328,125**
		11		**.343,75**
			23	**.359,375**
3				**.375**
			25	**.390,625**
		13		**.406,25**
			27	**.421,875**
	7			**.437,5**
			29	**.453,125**
		15		**.468,75**
			31	**.484,375**
1/2				**.500**

FRACTIONS				DECIMAL EQUIV.
8	16	32	64	
			33	**.515,625**
		17		**.531,25**
			35	**.546,875**
	9			**.562,5**
			37	**.578,125**
		19		**.593,75**
			39	**.609,375**
5				**.625**
			41	**.640,625**
		21		**.656,25**
			43	**.671,875**
	11			**.687,5**
			45	**.703,125**
		23		**.718,75**
			47	**.734,375**
3/4				**.750**
			49	**.765,625**
		25		**.781,25**
			51	**.796,875**
	13			**.812,5**
			53	**.828,125**
		27		**.843,75**
			55	**.859,375**
7				**.875**
			57	**.890,625**
		29		**.906,25**
			59	**.921,875**
	15			**.937,5**
			61	**.953,125**
		31		**.968,75**
			63	**.984,375**
1				**1.000**

THE BULLSEYE MIXTURE
by
Guy Lautard

This story is dedicated to Jim Davidson, a very fine old gentleman whose friendship I was privileged to share, who taught me much, and whose memory I treasure. He was not a machinist, but he was a man with many interests, and excelled in most of them: shooting, rose growing, photography, gem cutting, lapidary, ham radio, and more. In these and others he found pleasure and many friends until his death at 87 in 1975.

I hope that all who read this story will enjoy it, and perhaps learn something of technical value. I hope that you will also find in it things less tangible, but also of value.

NOTE: Any resemblance between any character in this story and any person living or dead is unintentional and purely co-incidental. The only exception to this is "Sam", who is in many ways like my old friend Jim Davidson.

THE BULLSEYE MIXTURE

by

Guy Lautard

Chapter 1
END OF AN ERA, AND A NEW PROJECT

It was almost 4:30 on a cool, late September Sunday afternoon when Dave Thomas cleared the Fort Lewis Army Base main post gate and fed his big Olds into the northbound traffic on Interstate 5.

In the passenger seat beside him, Sam Carter looked tired but happy. His day had started somewhat before 5 that morning, when Dave had picked him up at his home in West Vancouver, north of the 49th Parallel, for the 200 mile drive to Ft. Lewis, south of Seattle, Washington. Sam had shot in 3 matches during the day. He'd placed first in a 1000 yard service rifle match, cleaning up on about 200 other shooters in his class. He and Dave had been on the 4-man rifle team that had placed first in the 1000 yard team match. It had been a long day for a man turned 88 last month.

"How you doin'?" Dave asked.

"Not so bad for a young fella," Sam smiled. Even his voice was tired, but his smile was the same-as-always smile that folded his face in all the familiar seams - seams put there by sadness and laughter and concentration. The sadness was for the loss of his son, and the more recent loss of his wife of nearly 65 years. The laughter was from his innate good nature, and his many friendships. The concentration was from long practice of his trade: - machinist, tool maker, die sinker.

"You want to get some supper in Tacoma, or do you want to wait and get something to eat further north?" Dave asked.

"Let's eat somewhere up around Mt. Vernon."

They drove in silence for a few miles, then Sam gave a happy chuckle. "We did pretty good today, didn't we, Dave?"

"We did that," Dave grinned. "The team match was the best, though. Man, when Bill Duffy and Steve Coates saw what you did with that old Number 4 of yours..."

"Yeah - and when you put up a score almost exactly the same, they couldn't dump their rifles fast enough. Just think, Dave, 4 men, all shooting the same rifle in turn, and we cleaned up the whole line." Again the familiar smile, and the chuckle.

"I wonder if that's ever happened before?"

"I don't know. I've never heard of it, and I've been in the game since 1904, when I was 14."

They had met as shooters 5 years ago, in 1973 and soon learned that they were also nearly neighbors. Due to common interests, a close friendship had rapidly developed, and Sam had offered to teach him everything he could about the machinist's trade. Dave had stood in open-mouthed astonishment when he'd first seen Sam's basement shop. A 12" x 48" Sebastian lathe, a Hardinge universal mill, and a 7" South Bend shaper were the mainstays, along with everything one might want in the way of hand and precision tools. Dave had always been interested in metalwork, but his education and work had led him down different roads, and it was not till he met San Carter that he realized that his interest in metalwork might, with effort and dedication, be translated into a working knowledge.

Sam had "adopted" him as an apprentice, and they had spent countless hours together in Sam's basement. Although Dave had no illusions that he could hold his own for speed in a commercial machine shop, he had become a rather skilled machinist within certain spheres, due to Sam's instruction, plus his own intense desire to acquire the skills of the trade.

"I've been thinking," Sam said, breaking into his reverie, "it's about time for me to give up this long range competition shooting."

"Oh, come on, Sam - you'll be shooting the pants off guys my age for years yet!"

"I don't think so. I really felt like an old man today. I think I'll make this my last shoot - go out in a blaze of glory, so to speak. As of today, the Number 4 Lee Enfield is your rifle, Dave."

Dave's head turned sharply towards the old man. "You can't just give..."

Sam laughed. "Davy, my boy, the #4 belongs to me, and I can do with it as I please. I've shot four barrels out of it in the last 20 years. I don't want to hang it up in my basement and never see it used again."

"But...."

"Listen, son, that rifle doesn't own me - I own it. And it'll give me more pleasure in your hands, being used, than it will gathering dust in my basement. I want you to have it."

"Well, if you're sure that's what you want to do with it, I'm mighty proud to think you'd want me to have it. I don't know what I can do to thank you."

"You don't need to do anything to thank me. But I have a little project I've been thinking about for some time now, and I could use a hand with it, if you're interested. It's something I've wanted to do for a lot of years. Let an old man tell you a story, to give you some background, okay?"

"Okay, story away."

"My grandfather was a foreman with the Sharps Rifle Co. when they were making the Model 1878 Sharps Borchardt rifles. When the company went out of business in 1881, he was given two completed Sharps Borchardt actions, all finished, polished and ready for use. He built a rifle on one of those two actions. That rifle eventually fell into the hands of his oldest son, who was my father's brother - my uncle. There was bad blood between him and my dad, and I never met him. So that accounts for one of the actions, which I suppose is in the hands of somebody descended from my uncle, if it didn't get sold out of his family somewhere along the way.

"My father got the other action when my grandfather died in 1893, a couple of years before we moved from Bridgeport, Connecticut to Canada.

"Well, my father never did anything with that Sharps Borchardt action. Besides me, there were 3 sisters in our family - I was the only boy. When I completed my apprenticeship as a tool and die maker back in 1911, my father gave the Sharps action to me.

"Somehow, over the years, I mislaid it, but I found it about the time I retired, and I thought then that I'd like to build a rifle on it someday. That was about 1955. A single shot rifle was useless in military match rifle shooting, which was my main game then, so I never got around to doing anything with it. It's just been sitting there in the basement, wrapped up in greased rags from then till now. I dug it out yesterday."

"A factory-new Sharps Borchardt action'd be worth a bundle today!"

"Maybe it would, but that don't matter to me. How'd you like to help me build a rifle on that action?"

"Sure. What sort of a rifle do you want to make?"

"Well, let's talk about that for a minute. The action's not exactly ideal for a target rifle.

What about building a practical hunting rifle?"

"Sounds good," Dave agreed. "You could shoot it from a benchrest out at the rifle range. And we could go chuck hunting up country some weekends next summer."

If Sam didn't want to pursue the long range competition game, Dave was quite happy to bend his own interests to suit. His friendship with the old man meant far more to him than the particular form of their shooting activities... regardless of what Nora might think.

"What caliber would you want to build it in? Myself, I've always favored factory calibers, but maybe you'd like to make something unusual - a 'wildcat'?"

"Now just a minute, Sam. It's going to be your rifle, not mine. I'm just going to help you build it."

"Not quite, Dave. I want you to help me build it for two reasons. First because I'm not as good on the lathe as I once was, and second, because I want you to have the finished rifle when I pass on."

"But you just gave me your #4 and..."

"The #4 was mine to do with as I wanted, as I already told you, Dave. And what I wanted to do was to give it to you. Now, if you want to, you can help me do something I've always wanted to do. And when I'm done, the rifle we build will be yours."

"But what about Oscar..."

"Don't worry about Oscar. He's going to get the other three rifles - the Model 70 Winchester Bull Gun, the Remington 40X Rangemaster, and the Model 52 Winchester. But I want you to have the Lee Enfield and the Sharps. Oscar's got no interest in the #4, and he certainly won't have any stake in the Sharps if you and I make it."

"Well, that's mighty nice, for me, Sam. But if the Sharps will ultimately be mine, then I want to pay for the barrel and the stock and any other items that go into it."

"Okay, that's fair, if you want to do it that way. Let's talk about what'll go into it, and so on."

"Okay. I don't think we could go wrong with a Douglas premium barrel, to start with - a good, medium weight sporter blank, about 26 inches long. We'll carve the stock from a rough blank, so we'll just see who's got some really nice walnut, or whatever. You got any preferences on the type of wood we use?"

"No. But you said last week you'd been looking at Alvin Lindin's book *Restocking a Rifle* and you kinda liked the look of his "sugi finished" maple stocks. You want to do it in maple?"

"Yeah, lets' do that!"

"What about the buttplate? You want a recoil pad, or a steel buttplate?" Sam asked.

"What about a color casehardened steel buttplate? You've told me more than once you'd teach me about casehardening, but we've never got around to it."

"So I have. Well, this'd be as good an excuse as any. We'll do that. What about sights - you want to put a scope on it?"

"I've got that Leupold 3x-9x variable I bought from Alec Bullard last year."

"That'd be perfect." Sam heaved a sigh. "Davy, I'm a tired old man. Do you mind if I go to sleep for a while?"

"No - that's fine. I'll wake you about 6:30 and we'll stop for supper in Mt. Vernon. In the meantime, let's have a little music."

He pushed the "Wonderland By Night" cassette into the tape deck and relaxed in the quiet but compelling instrumental music. Sam liked Bert Kaempfert as much as Dave did. If he had not, Dave would have had tapes of whatever his old chum did like, and he'd have played that.

While Sam slept, Dave drove northward, cogitating on the events of the day. They'd had good luck at the matches, that was for sure. His thoughts turned to the fact that Sam's #4 Lee Enfield match rifle was now his - he felt honored that Sam would want him to be its next keeper.

To the unpracticed eye, it was a very ordinary rifle. Most shooters would see it as just another .303 British army rifle. Such was not the case. It was a match rifle, and any shooter who knew the Canadian Service Rifle match game would place it as such immediately by the blocky Parker Hale 5C rear sight. Some would also note the Lyman globe target front sight tucked neatly in between the modified "ears" which protect a Lee Enfield's simple blade type battle sight. Only close examination would reveal the fact that the rifle was not chambered for the usual .303 British cartridge. The first 4 barrels had been for that caliber, but on its last trip to the Long Branch Arsenal in Winnipeg, Sam had asked that a match-grade 7.62 NATO barrel be installed.

Most Lee Enfields require the addition of a reinforcement, in the form of an ugly, welded-in steel plate on the receiver, when this conversion to the higher pressure 7.62 NATO round is made. Sam's #4 had not, and the rifle, which to most eyes is not particularly handsome in any case, was the better for the avoidance thereof. The caliber change had cost Sam the use of the magazine, but that mattered naught in the game he favored.

Now it was to be his. What a keepsake of his friendship with Sam! As for the Sharps Borchardt - Dave did not particularly desire to own such a rifle, but again, if it so pleased Sam that they should build it together, and that it should one day be his, well, that was very nice... But he'd rather have Sam alive and well and active than 20 such rifles, if the choice were his.

But both he and Nora had begun to notice small signs of decline in Sam in the past few months, and both realized he would not be with them many more years. Dave had long since realized that so long as he might live, he would never meet another man as fine as Sam Carter.

The music changed to a new tune. Sam settled deeper into his seat. Dave drove on, part of his mind on the road and the traffic around him, part of it swinging onto the matter of Sam's granddaughter, Nora.

Sam's son, Duncan, had married in 1943, while on leave from the Navy. From that union Nora had been born, but her father saw her only once, when she was about a month old. Three weeks later, he was killed when the Royal Canadian Navy corvette HMCS *Shawinigan* was torpedoed and sunk in Cabot Strait, with the loss of all hands, in November of 1944. Little Nora and her mother had

moved in with Sam Carter and his wife - accommodation was hard to find in wartime Vancouver. Nora's mother died before she was two, and Nora had been raised in her grandparents' home. She had been born with a dislocated hip which had gone undetected until she was nearly two. In spite of intensive therapy, surgery and treatment extending over the next 5 years, and paid for at enormous sacrifice by her grandparents, she walked with a very pronounced limp, and always would.

She was a skilled calligrapher, as well as a talented artist with a pencil, and made a nice living for herself from a small, bright studio in her grandfather's home.

Dave had met her when Sam first invited him to his home. The first thing he had been struck by had been her poise and attractive appearance. The second thing he had noticed was her limp, when she came across the kitchen to shake hands with him.

No one lives with such a handicap without becoming highly sensitive to other peoples' reactions to it. Nora had immediately sensed his disinterest in her as a woman and as a person - a disinterest he, with shame and sadness now had to admit he had felt, seeing her first three or four steps across the room.

She had also sensed his early attitude towards her grandfather's equipment, both the shooting gear and the machinist's tools. To put no fine point on it, his first feelings had lain somewhere between "envious" & "covetous". This too Nora had perceived, and she disliked him for both.

Within half a dozen visits to the Carter home, he had ceased to notice Nora's limp, as her fine character and winning personality became evident. Within 2 months he would happily have married her, but she remained aloof from anything beyond the courtesy she would extend to any visitor to her grandfather's home.

Not that he had failed to show his regard for her. She had had difficulty getting into his low-slung sports car one day when he offered to drive her somewhere. Within the week he had sold the car and bought a 4-door sedan - a used '67 Olds in beautiful condition, and with very low mileage on it. Sam had not been entirely pleased, at first, for he liked the sports car, but Dave had explained: "I thought it'd be easier for Nora, if we ever go any place together - the three of us." At that, Sam had nodded, and decided that the "Old-mobile", as he and Dave came to call it, was fine.

But it had not softened her heart. Her aloofness stemmed not from what he did, but from how he had first measured her. That he had held her in the highest regard for 59 out of the 60 months he had known her moved her not an inch.

His covetousness towards Sam's gear had died more slowly, under the influence of the old man's friendship and the values by which he lived. Dave had known his feelings for what they were from the outset, and had not been proud of them, but he found it difficult to feel otherwise. Sam's shop equipment was first class in every respect, from the three major machine tools down to the least center-punch - Starrett, Brown & Sharpe and Lufkin were the makers of almost every tool in his three toolboxes, with the exception of those he had made himself.

Whether Sam had ever suspected his early attitudes or not, Dave did not know. From the outset, Sam had been friendly, helpful, and generous with his time, his knowledge, and his workshop. Dave was genuinely and intensely interested in learning about metal-work and the skills of the machinist - that had certainly been evident. Sam had taken him at face value, and they had become friends, chums, pals and partners in many a workshop project.

Sam had started him on the making of a set of basic machinist's tools: firm joint calipers, V-blocks, machinist's squares, and so on, teaching him the trade he himself had prac-ticed with consummate skill for nearly 70 years. In the 5 years since Dave's "apprenticeship" began, he had acquired a creditable measure of his mentor's skill and knowledge, and all of his love of the trade.

Dave Thomas took the 2nd exit for Mt. Vernon, and began talking to Sam to wake him. They stopped at a restaurant of Sam's choosing and had supper. Dave consumed a full meal, followed by an ice cream sundae as big as a football. He offered to bet the price of the entire meal with the waitress, who laughingly expressed her doubt of his ability

to finish the sundae. Sam ordered a light sandwich followed by a cup of coffee.

"You shouldn't drink that stuff, Sam. It'll kill you," Dave teased.

"Oh, no. This is the Bullseye Mixture!"

It was a never-varying exchange whenever the two of them ate together.

Anything Sam approved of was "The Bullseye Mixture."

Chapter 2
HOMECOMING

"What's the matter with your friend?" the Canadian Customs officer asked when Dave pulled into the Peace Arch border crossing just north of Blaine.

"Nothing. He's had a long day. He's 88 years old, and if you can leave him to sleep, I'd appreciate it."

"Okay - he looks pretty harmless. Have either of you got anything to declare?"

Dave felt like telling him that the old fella could put a hole in his hat at a thousand yards with iron sights, but refrained.

"No. We were down to a rifle match at Fort Lewis. Our rifles are in the trunk. You want to see 'em?" Dave asked, taking a green ticket from his shirt pocket and handing it to the Customs officer.

"Yes, please. Don't tell me that old geezer can shoot?"

"You better believe he can, Mister. There were a lot of guys at the matches today wished they could shoot half as well."

He got out, opened the trunk, and uncased each rifle in turn. He was soon on his way again, and an hour later pulled up at Sam's door.

"Wake up, Sam. We're home. Nora's got the porch light on for us."

Sam roused himself, and they got out of the car. Dave took Sam's shooting box and the #4 from the trunk, and they started towards the front door. Half way up the walk, Sam stopped. "What are you bringing the #4 in for, Dave? Take it home, clean it, and put it in the guncase with your 40X."

"Well, I ... are you sure you want me to have it, Sam? I..."

"Yes, I'm sure. You think I've lost my marbles? Take it home, and make an old man happy."

They stood facing each other in the path, in the yellow light of the porch bulb, and a volume of unspoken feeling passed between them. Dave returned to his car and put the #4 in the trunk. He put his free arm around Sam's shoulder as they went up the steps together.

"We had fun, didn't we?"

"So we did, Davy, so we did." The door swung open then, and Nora stood aside to let them in, greeting them as she did so.

"Hello, Lass. How's my girl tonight?"

"Just fine, Grampa. How was the shoot?"

"Never better. Davy and I cleaned house." His smile was enough for her - if he was happy, she was content.

"Hi, Nora," Dave said quietly, closing the door with his shoulder.

"Hi. Come on through to the kitchen, both of you. There's coffee on, and some pie. You can tell me how you cleaned house while you eat."

"I'll just put this stuff downstairs, and I'll be right up," Dave said, opening the door to the basement stairway.

"Thanks, Dave. You'll find a package on the workbench, wrapped up in a piece of oilcloth. Bring it up when you come, please."

As he put Sam's gear away, he could hear Nora setting the table for their homecoming snack. There would be no coffee cup, but a glass of ice-cold milk at his place, he knew.

The oilcloth-wrapped package was where Sam had said. He took it upstairs with him, and set it down beside Sam's elbow.

"Good. Sit down and have some pie. Best apple pies you'll ever find, made right here in Nora Carter's kitchen."

This statement was difficult to doubt. The apples in Nora's apple pies were grated into coarse shreds, and besides being spiced to perfection, somehow ended up almost as firm in the pie as raw apples. How she did it she would not say.

After recounting their day's successes for Nora, Sam asked what had gone on at home while he'd been away.

"There were three phone calls. A lady in Burnaby phoned to ask you something about her rose bushes, and a man in White Rock called to ask if you'd cut up an opal for him, and another chap wanted your advice on buying some gem cutting equipment. I got all their numbers, and told them you'd call back in a day or two. And Pete came over to mess around with your lapidary equipment."

"Good. The usual complaints, I suppose?" Sam chuckled.

"Yes. 'Always the stones they come crooket when I do chusta like Sammy say. I get Sammy to fixet me up next week.'" They laughed at Nora's imitation of Pete. He was a retired storekeeper with an unpronounceable surname, about 20 years Sam's junior, who lived in a nearby apartment building. He had met Sam at a lapidary show, and had been invited to use Sam's rock polishing equipment. He came several times a week, his enthusiasm undiminished by the less-than-perfect results of his heavy-handed approach to his chosen hobby. Usually he would reach a point where he would have to ask Sam to correct the shape of the stone. He would watch with glee and amazement as Sam's masterful hand would in minutes right the shapes he had bungled over for hours.

When Dave had finished his pie, Sam indicated the oilcloth-wrapped package with his coffee cup. "Open that up, Dave and let's have a look at it."

Dave dug a pocket knife from his jeans, and cut the strings on the package. The oilcloth was very old, and crumbled into pieces when he began to remove it. He looked at Nora apologetically: "I'm making a mess of your table. I'll get a piece of newspaper..."

"Just sit still a minute - I'll get you some." She went into the living room and came back with a section of newspaper. Dave set his pie plate on the counter while she spread the newspaper on the table for him.

"Thanks Nora - that's great." Under the oilcloth was a wrapping of cosmoline-soaked cotton rags, which came away easily, exposing the item within.

"Look at that!" Dave breathed. "A mint 1878 Sharps Borchardt action."

"Were did that come from, Grampa?"

Dave passed it across the table to her. She examined it in the light of her grandfather's explanation. "So this belonged to my great great grandfather. Isn't that neat!? And you two are going to build a rifle on it?"

"Yes. And just so there's no misunderstanding later on, in case Oscar or anyone else ever raises the matter, Dave is paying for the barrel and the wood for the stock. We'll be using one of his scopes on it. And he'll do most of the work of building it. I told him it would eventually pass into his hands, when I asked him this afternoon to help me build it, and he said that in that case he would buy the necessary parts."

Nora glanced briefly from her grandfather to Dave, and nodded.

A few minutes later Dave left the Carter home, having agreed to return the following evening. His own house was only a few minutes drive from Sam's. By the time he had carried the last of his own gear in, it was a little after 10:30. He had been up since 4 AM, driven something over 400 miles, and shot in a series of highly competitive matches during the day. But he cleaned his 40X and Sam's old #4 Lee Enfield match rifle thoroughly, and locked them both in his gun cabinet before turning in for the night.

He dreamt of Oscar, who tried to wrestle the Lee Enfield from his hands, then turned into a leering figure in a medical gown, saying, "So what if she's got a crippled hip. Who cares about her?" He woke up, his fists opening and closing like hydraulic vises.

Chapter 3
PICKING THE MAKINGS

Monday evening Dave dropped around to see Sam. Sam let him in at the basement door, and led him through the various "departments" - photographic film processing, printmaking, lapidary, gem cutting, etc., to the main workshop - "the shop", as they referred to it - where stood the lathe, the mill, the shaper, and a workbench. On the latter sat the Sharps Borchardt action, fully cleaned up now, and reposing on a piece of navy blue felt. Nora would have supplied that.

"Well, did you get the lady with the roses and the opal man fixed up?"

"Oh, yes. And the fellow who wanted to know about gem cutting equipment. The opal man is coming to see me tomorrow. And Pete was here this afternoon. Are you about ready to order a barrel?"

"Yes. If we can decide on what bore size tonight, I'll phone Douglas tomorrow and see what kind of delivery dates we're lookin' at."

"And wood?"

"That shouldn't be a problem. I'll call one of the hardwood places right here in town - if they don't have what we want on hand, they can get it, or tell me where we can get it. Would you prefer birdseye or fiddleback maple?"

Sam shrugged. "Fiddleback, I guess. Seems more... more traditional, somehow. What do you think?"

"The same. I was hopin' you'd say fiddleback, and not birdseye," Dave admitted. "What about the buttplate - you done any more thinkin' about that?"

"Yes. How about we shape one up from a piece of sheet steel, and color caseharden it, like you said Sunday?"

"Sheet steel? Isn't that going to look pretty crude?"

"Not the way I have in mind to do it. We can use a piece about 3/32 thick, and form it to a nice shape, about like a Biesen..."

"How do we do that?"

"By careful hammering over a rounded steel block."

"But then it'll have hammer marks all over it."

Sam Carter smiled his wrinkled old smile, and he laughed that soft laugh that you'd think was almost a cough, if you were hearing it for the first time.

"Not when we get done with it, Davy. We'll file all the hammer marks out of it, and we'll file all the edges true, after we've got it blocked out the way we want it. And then we'll polish it."

"But if it's polished, it'll slip on your shoulder..."

"No. We'll stipple it with a punch, before we polish it. The punch will throw up little teeth all over, and the polishing will smooth them down just enough to make it non-slip."

"You think of everything, don't you?" Dave grinned. "You said we could caseharden it. Can we do a really good job of that ourselves?"

"We'll do it ourselves, and do as nice a job as any commercial shop'd do for us," Sam assured him.

"With cyanide?"

"No. We'll do it the old way, in a carbon pack. That's how this was done..." He handed Dave a toolmaker's surface gage. The sides and top of the rectangular base exhibited a beautiful play of mottled blues, reds, straw yellows, purples, browns and greys. The bottom face had been lapped to an almost mirror finish once, dulled now by a lifetime's service to its maker: "SWC - 1908" was neatly stamped on one end.

"You made that in 1908? You'd have been.... 18 then, right?"

"Yes. I made every piece, made my own case-hardening compound, packed that base in it, and put it in the forge for a couple of hours. The case on that would be about 20 thou deep."

"And we can do our buttplate the same way?"

"Sure, though half that depth of case will be plenty for a buttplate. I'll tell you more about how it's done tomorrow night, if you want to drop over."

"Sure. I'd like to. You feel like thinkin' about what caliber to chamber the Sharps for tonight, or not?"

"Sure. Let's pull out some ballistics tables..."

They argued and discussed and compared for half an hour, at the end of which time they had decided on a .30 caliber bore, and that they would chamber it for either the .30/30 or .30/40 Krag, both these being rimmed cases well adapted to the Sharps Borchardt extractor. They chose the .30 caliber bore because between them they had 8 molds for 5 different cast bullets in that size. Ultimately, they decided on the .30/40 Krag cartridge.

Finally they heard Nora in the kitchen, a sure sign their evening was about to come to an end over some sort of snack.

Later, when Dave left, Nora contrived to see him out to his car.

"Dave, I wanted to speak to you. Grampa was awfully tired today from the trip to Ft. Lewis yesterday. And he was sick during the night - throwing up, I mean. I'm really worried about him."

He frowned. "When's the last time he saw a doctor?"

She rolled her eyes. "40 or 50 years ago. He passed a kidney stone then, and he's been as healthy as a horse ever since. Or so he says. I'd like to get him to go to a doctor, but I don't think he would."

"Would you like me to talk to him?"

"Would you, if I can't get anywhere with him?"

"Of course. I'll do anything for him. You have my phone number at work, don't you? Well, if anything comes up and you need any help, and I mean ANY kind of help, you call me, at home or at work. If it's 2 o'clock in the morning, or whatever, just phone me - I won't mind at all. Okay?"

"Okay. And thank you, Dave."

"You don't need to thank me. I have 2 favorite people in this world, and Sam Carter is one of them."

"Who's the other one?" She asked the question quickly, without thinking, and as quickly wished she'd had wit sufficient not to have done so.

He looked at her for some seconds, debating his answer, then said simply, "You are."

She made no reply.

"I'd best be going. Thanks for the snack. And thanks for telling me about Sam. I'll talk to him about seeing a doctor if you can't get anywhere with him. Call me at work tomorrow, when he's downstairs, and let me know what luck you have with him, okay?"

"I will. And thanks. Good night."

Chapter 4
A START ON THE BUTTPLATE

At six o'clock the next morning Dave Thomas phoned Charleston, West Virginia to order a .30 cal. Douglas barrel blank with a rifling twist of 1 turn in 12 inches. Later that morning, from work, he made several calls to local hardwood dealers, but failed to turn up any fiddleback eastern rock maple. However, one agreed to try to locate some, and said he'd call him back if he was able to do so.

Nora called him about 11: Sam had been sick again overnight, and surprisingly, had agreed to see her doctor. The appointment was for that afternoon.

"They want to do a bunch of tests on me," Sam growled that evening, laying tones of special disgust on the word 'tests,' as he told Dave about having seen the doctor.

"Well, that's good. Maybe they'll find out what's making you sick, and then fix you up better'n new."

"How did you make out about a barrel?" Sam asked, changing the subject. Dave told him, and about trying to locate the wood for the stock.

Sam handed him a half-finished buttplate. "What do you think of this?" Although it was not done yet, it was already recognizable for what it was to become.

"You'll have some filing to do in a few minutes," Sam said, taking the part back again.

As Dave watched, the heel of the buttplate began to take shape under Sam's hammer.

They both knew a highly satisfactory checkered steel buttplate could be purchased for perhaps $10. However, on a project such as this, the making of certain parts from scratch was one of their satisfactions. For Dave it was a total relaxation from his job, and an opportunity to learn. For them both, it was an excuse to spend time together, sometimes silent for long periods, sometimes discussing matters pertaining to the project at hand, or far removed from it.

Dave watched as Sam's sure hammer blows coaxed the metal into the shape he wanted. After a while Sam put the buttplate between the padded jaws of the vise. "How'd you like to file that a little closer to the final outline?" he suggested.

Dave picked a file from a rack containing at least 40 files, and proceeded to address himself to Sam's unfinished creation.

As he worked, he recalled how he'd once stood at this same vise almost every evening for several weeks, filing a piece of 5/8" hot rolled mild steel bar into a hexagonal plug under Sam's patient tutelage. Eventually the plug, if oiled, could be pressed with heavy thumb pressure through a hexagonal hole in a piece of 1/4" thick gage plate, in all six orientations.

"You're not the first apprentice I've had make a plug like that, Dave, but you'll be the last." Sam had said, picking up a small hammer, and putting a centerpunch mark on the edge of his "master." There were 27 punch marks - one for every apprentice's plug that'd gone through.

The hex plug itself was of no value, but making it had taught Dave what can be done with a file by a man who has learned to use one properly. He knew he wasn't as good with a file as Sam - Sam had filed out the hex hole! - but he also knew he was better than many professional machinists, and he took considerable pride in his skill.

Eventually he stopped filing. "I don't want to go much further till we get the buttstock close to size, and then we can tailor the one to the other as we go."

"Good idea."

"You were going to tell me all about case-hardening tonight," Dave reminded him.

Sam's watery grey eyes smiled at him through his heavy glasses. "Take more'n one evening. Do you know much about it yourself?"

"Not very much. Why don't you give me the full treatment - why it's done, and where, and how?"

Chapter 5
A LECTURE ON CASEHARDENING

"Well, my voice won't stand telling you the whole story at one go, but let's start with why it's done. You interrupt me if you want to ask anything."

Sam settled himself on his "talkin' stool". Dave Thomas leaned against the old Sebastian lathe.

"You know," Sam began, "all about hardening high carbon steels like drill rod, old files, and so on..."

Dave nodded.

"And you know that low carbon steels can't be hardened by the same methods?"

Dave nodded again.

"Do you know why not?"

"Because there's not enough carbon in them for them to harden when they're quenched from above the upper critical temperature, which is where the carbon goes into solution in the steel, and the steel becomes non-magnetic."

"That's right. Your average piece of low carbon steel, the so-called mild steels, will have from about 0.1 to, say, 0.3% carbon. Hardenability starts at about 0.4% carbon content. 4140 - which is a chrome-moly steel, not a mild steel - has a carbon content of about 0.4%, and it'll harden. It isn't an ideal tool steel, but it can be made to serve in some materials, and/or where you don't need long cutter life. Anyway, mild steels do harden in a couple of circumstances, and they can be a bit of a nuisance for it, too.

"For example, if you were to take a piece of ordinary cold rolled steel, which would have say 0.2% carbon in it, and cut it with an abrasive cutoff saw, which'll go through it like a red hot knife through butter, the cut surfaces will be so hard you can't touch them with a scriber. That's one instance. The other is flame cutting. You cut a piece of hot rolled mild steel plate with an oxy-acetylene cutting torch, the cut edge may be so hard

you can't readily machine it with a high speed steel tool. Now then, do you know why?"

"No, but I ran into that one time, and you told me to grind the flame-cut edge away on the bench grinder, and after I got through the hard stuff, I put it back in the lathe and away I went."

"I remember that. What happens is that the high localized heat draws carbon from the surrounding steel towards the cut faces, which increases the carbon content. The mass of steel behind the cut draws off the heat rapidly enough to act about like a quench, and the result is a hardened surface."

"That's in the case of the flame cut metal. What about the abrasive cutoff saw - what happens there?" Dave asked.

"Same thing. The abrasive saw goes through the steel very quickly, but you can bet the temperature right in the cut skyrockets, and then drops off, probably faster than when a cutting torch is used, because the **amount** of heat supplied by the torch is so much more. You look at the end of a piece of steel cut with an abrasive saw some time, and you'll see there's been some hot doin's there! And if you try to scratch it, don't be surprised if it's too hard for anything 'cept a carbide scriber."

Dave nodded thoughtfully, taking in what Sam had said.

"Well, that's a couple of instances of when mild steel **does** harden. Now then, I started out to tell you why mild steel is sometimes casehardened. Low carbon steels, being low in carbon, can't be hardened overall like high carbon steels, by heating and quenching. Being soft, though, they are not liable to fracture under shock loading, where a piece of hardened tool steel, being not only hard, but also relatively brittle, would shatter. Try using a file as a pry bar and it'll snap like a carrot."

Sam shifted slightly on his stool and continued: "A steel gear is a good example of where casehardening might be used. The gear teeth are subject to shock loads when the mechanism slows down or speeds up. So the gear has to be tough, to resist fracture. At the same time, it has to be hard, to give long service life. So it may be made from soft, low carbon steel, for fracture resistance, and then casehardened to give it a hard, wear-resistant skin, or "case." Rock drills are case-hardened - soft to stand the shock, and casehardened against wear. The inserted high impact steel cutting edge, if there is one, will be totally backed up by the drill steel, so it won't fracture, and break up in use."

"But how...?"

"Don't worry about how for a minute, Dave - I'm just explainin' the why part just now. You know casehardening is employed fairly widely in guns. Why do you suppose that is?"

"Well, from what you've just been saying, I suppose so the repeated shock of firing won't eventually fracture the frame, or the hammer, or whatever other parts there are that are subjected to shock loads, yet they'll also stand up to the wear of one part rubbing against another."

"That's mostly right. Soft steels are also easier to forge and machine, so if you can make a gun part - a frame, say, or like you mentioned, a hammer - from soft, low carbon steel, it'll be easy to work. Easy on your cutter, easy to file if there's any hand fitting to be done - and when it's all done, you case-harden it. Now the gun maker may do his casehardening in such a way that the part comes out with not only a hard skin, but a nice play of colors on the surface, too. That's commonly called "color casehardening," and as you know, your average gun nut goes crazy over a good color casehardening job. He may not know how it was done, or why, but he likes the look of it."

"Yeah! Colt's Single Action Army frames, and Smith & Wesson hammers.. Okay, I can see why they'd be casehardened," Dave said. "But you said the casehardening gives the soft steel a hard skin. You don't mean on the outside, in the sense of *adding* hard steel over the part? That'd change the dimensions...?"

"No. The outer steel of the part itself is transformed from a low carbon steel to a high carbon, heat-treatable steel," Sam explained.

"You mean they heat the part up enough to make carbon in the interior of the part migrate to the surface?"

"No, I didn't say that. The work is immersed in something rich in carbon - charcoal, for example, or molten cyanide salt - and heated.

When the work gets hot enough, say up about 16-1700°F, carbon will migrate *into* the steel from the surrounding casehardening medium. Typically, we will bring the carbon content of the surface layer - the "case", as it is called - up to maybe 1 to 1.2% carbon, while the "core" will still be what it was to start with: low carbon steel - in other words, steel with a carbon content of say 0.2% carbon."

"Now that I've explained this much to you, I'm going to backtrack on myself, and divide what I've been calling "casehardening" into 2 separate processes - 'carburizing' and 'heat treating.'"

"Carburizing," Sam said, raising one finger, "is the addition of carbon to the surface of a piece of low carbon steel. Whether that surface layer ends up hard or not, as well as the condition of the core, depends upon (here Sam raised the other finger) the subsequent heat treatment. If you don't heat treat it in such a way as to harden it, you will still have a piece of soft steel; it just happens that the surface layer has more carbon in it now. Are you still with me?" Sam asked, shifting himself slightly on his stool.

"Yes, I follow you. I'm just wondering if the next thing you're going to tell me is that the term "casehardening" is a complete misnomer."

"No! 'Casehardening' is a very handy term, and it simply means the complete process of carburizing and hardening. I wanted to explain what I've just told you so that when you read the term "carburizing" in something - in *Machinery's Handbook*, for instance - you will realize that it is not some *other* process, it is part of the casehardening process. And also so that you will understand that you don't simply caseharden something; you get the surface layer's carbon content arranged so that it *can* be hardened, and then you heat treat it so it *is* hard. You may also want to affect the condition of the steel in the core, which is why you may heat the work twice, or even three times: once to carburize the surface layer, then either a quench to harden the case, or a slow cool, leaving the case soft, followed by re-heating to refine the grain structure of the core, followed by quenching to harden the case, and possibly a third heating to temper the dead hard condition of the surface layer."

"I love it! You old people really know how to explain things!" Dave said with a grin. He often teased Sam by saying "..you old people..", but it was a comment filled with respect and affection, not otherwise.

Sam patted his snow white thatch and chuckled, "It's this white stuff that does it. Now then, to continue with your education: The longer you leave the work at heat, in the presence of the carbon supply - in other words, the longer the carburizing process is continued - the deeper the carbon will penetrate, and the deeper the case will be. If the workpiece has some quite thin sections, you want to be careful you don't drive the case in so deep that the part is hard right through in the thin area - if you do, it will likely fracture or snap right off at that point when you put it in service."

"What kind of depths are we talking about? You said last night the case depth on your little surface gage was about 20 thou."

"Well, like I said, it depends on how long the carburizing operation is allowed to go on. The case can be anywhere from a couple of thou to 3/8" deep, depending on the job. A big gear - say in a piece of heavy earth-moving equipment - would be an example of where a really deep case would be wanted. Other times, a job will be rough machined oversize, deeply casehardened, and then ground to final dimensions. You need a deep case here so that even after you reach your final sizes, you still haven't gone through the case. Plus, by grinding after hardening, you can correct any warpage that may have occurred."

Sam took a deep breath. "That's enough for tonight, Dave. You ruminate on that, and tomorrow I'll tell you some of the details of how it's done generally, and how we can do it ourselves."

"Okay. I'll clean up."

"No - just leave it. I'll be down here tomorrow morning anyway, and there's not much mess from what I've been doing today."

Dave ran a finger over the edge of the unfinished buttplate. "It's gonna be nice."

"Oh, I meant to tell you - Oscar was here this morning."

"Was he? I suppose he told you how to finish this off?" Dave said, indicating the buttplate.

Sam shook his head. "I didn't bring him down here. I figured we'd just do the Sharps without his advice. You'd prefer that, wouldn't you?"

"Yeah, very much so. Thanks, Sam."

"Grampa? Will you and Dave be done soon? I've got some cookies just coming out of the oven." Nora's voice interrupted them on Sam's homemade intercom.

"We'll be up in a few minutes." He turned back to Dave. "Fella came today about his opal. You have to see that!" He went to the bench bearing his lapidary equipment, and returned with a piece of rock half the size of his fist. "Look at that! Raw opal!"

Dave whistled an exclamation. "Where did that come from?!"

"From the Andamuka mines in Australia. The fella said his brother sent it up here with instructions for him to get it made up into several pieces of jewellery for their mother."

"And you're going to do it?"

"Well, I told him it was too big a piece, and too nice, to just cut up for small items of jewellery, but he said he didn't care, that's what he wanted done with it. So I told him I'd do it for what's left over when I've made all the things he wants - earrings, pendants, rings, and so on. There'll be enough left over for a king's ransom!" Sam chuckled, setting the opal down on the bench. "Nora came down and practically threw him out, to take me to the doctor."

"And she'll be throwing me out, too, if we don't come upstairs pretty quick."

He stayed only briefly upstairs, long enough to have a couple of cookies. Nora's jumbo raisin cookies were almost addictive. Sam suggested he come back the following evening to discuss some more about casehardening. Nora made no move to speak to him privately as he left, only saying good night and giving him half a dozen of the cookies, wrapped in waxed paper, "...for your lunch tomorrow."

"Okay, Nora. And thanks." He waved the cookies at Sam.

Sam smiled. "I know: best cookies in the world. That I should have such talented grandchildren! See you tomorrow."

Dave Thomas thought about Oscar Matlin as he drove home. A long-time shooting friend of Sam Carter, he had recently retired from a long career as a flying instructor, finishing up as senior instructor with one of the major airlines, teaching younger pilots to handle the biggest passenger jets.

Matlin was a flyer of vast experience, and no doubt very good at what he did. The trouble was he could never stop "instructing," even if the subject was one he knew virtually nothing about. The classic case had been the one about gem cutting... Dave chuckled to himself, recalling that one. A lot of shooters didn't like Oscar because of his abrasive, know-it-all ways. In fact, many wouldn't even speak to him.

"I've been a friend to Oscar because Oscar needs a friend," Sam had once explained to Dave. That knowledge had made it easier for Dave to tolerate him, and he got on well enough with Oscar whenever they happened to meet at the range or at Sam's house.

Oscar annoyed him more often than not, but he wasn't all bad, and for the sake of Sam's friendship with him, Dave tried not to let Oscar get under his skin. He did know a few things, like the deer hunting tip he had passed on to Sam and Dave two years ago: if you are out hunting deer, and come across a tall dead snag - a tree that has had its top knocked off by lightning, say and if you will take a rock maybe the size of your head or a little bigger, and hold it against your stomach and bump it hard against the snag, the snag will emit from its top a sound that will act like a deer call.

Dave had tried it the following hunting season, high on a mountain above his uncle's ranch in the south central Interior of B.C., and after 5 or 6 bumps spaced maybe 10 seconds apart, had waited to see what might develop.

Within three minutes the biggest whitetail buck he'd ever seen had appeared out of the timber not 50 yards away, and stopped short, head high, nostrils flaring.

The buck had paused a second too long, and its horns now hung in the ranch house living room with several other magnificent racks taken by his grandfather and his uncle over the 80 years since his grandfather had homesteaded the rich Beaver Creek bottomlands and built the ranch now run by his uncle.

Chapter 6
MORE ON CASEHARDENING

Dave hurried through his supper the next night, then headed down to Sam Carter's place. Sam had the buttplate all but done, and mentioned to Dave the other items they would need to make or buy for the Sharps Borchardt rifle's completion: scope mounts, sling swivels, and a sling. They would also need a set of roughing and finishing chambering reamers. Dave agreed to order a .30-40 Krag chambering reamer from Brownell's the next day.

Sam didn't feel like working any more that evening, so he picked up his lecture on casehardening where he'd left off the previous evening.

"You said you'd tell me how it's done, and how we can do it ourselves," Dave prompted him.

"So I did." He paused to scratch his head. "The methods used today in industry are mostly best left to industry - they're set up with proper ventilation and so on to make sure nothing unpleasant comes from using molten cyanide and all like that there. However, that your education may be complete, I mention it. Cyanide has the chemical formula "CN", which means it is composed of carbon and nitrogen. Both are pretty well inert alone, but together they are deadly, and then some. Carbon, from cyanide in the molten state, will migrate from the molten salt bath into the steel parts suspended therein, and when the process has gone on long enough, the parts are removed, and cooled, either rapidly or slowly. If the cooling is rapid, the surface layer will become immediately hard, and may then require some tempering, same as any hardened steel will, depending on the intended end use.

"If the part is cooled slowly, the surface steel, even though its carbon content has been increased, will be soft, like a file that's been annealed. The heat treater will run the part back up to some suitable temperature later on, and quench it in water, or oil, or brine, depending on various factors."

"What would dictate his choice, and why wouldn't he just quench the part direct from the molten salt bath?"

"Well, the particular steel would mainly govern the type of quench. Then again the application might govern it. That usually comes from experience. When many pieces of the same part are made and used, experience is gained as to the best heat treatment procedure. As for why not quench direct from the casehardening bath, we'll come back to that, but it's mainly to do with the fineness of the grain structure in the core.

"Like I said, all that is best left to industry. We can go back to the old time pack casehardening methods they used years ago, and do an entirely satisfactory job, right in our own back yard."

"What about bone dust as a casehardening agent? I've read about that, but nobody seems to get very specific about where to get it or how to use it. Do you know anything about that?" Dave asked.

"Not firsthand. I understand you can use regular bone meal, such as you'd buy at a garden supply store, with very good results. Apparently you just use the bone meal as a pack case hardening agent, straight - you don't need to add any activators, or anything - but I've never tried it myself. We can make our own casehardening compound right from scratch - old Sam's Special Bullseye Mixture - basically high grade charcoal, crushed, screened, and dosed up with some extra goodies to make it work faster. I'll give you the recipe later. Maybe sometime you can do some experimenting with bone meal if you want to, based on what I'm going to tell you about pack casehardening in general."

"What about Kasenit?" Dave asked. "Is it any good?"

"I've never done but just a couple of pins with Kasenit, but there are a few things to know about that stuff. If you use it the way the makers say to, you can get a very hard case, sure enough, but it will not be very deep - maybe 2 or 3 thou thick. By way of comparison, most of the casehardened parts in a M1903 Springfield - the sear, the trigger, the bolt, the receiver, and so on - these would be casehardened to maybe 8 to 12 thou deep.

"Kasenit has its place, but if you want to put a good deep case on something, it's not the way to go. I've read that you can use it as a pack casehardening compound - just like I'm going to tell you how to do with charcoal - and apparently, if you do, you will get a good deep case. I suppose that's logical enough - you're just giving it more time to do much the same thing as when you heat the job and roll it around in Kasenit powder. But using Kasenit as a pack casehardening compound, I don't think you'd get the colors you might also want. I suppose a fella could throw a handful in with his charcoal - just for good luck, so to speak. But on the other hand, it costs money, and the charcoal is just about free, except for the activators, so why bother?"

"Now then," Sam said, "that's about enough for tonight. How'd you like to do something to earn your keep and all this golden wisdom I've been dispensing? Drill a couple of holes in that buttplate where you think you want them to be. There's the screws we'll use to attach it to the stock," he added, indicating two #10 wood screws on the bench.

Dave swabbed machinist's blue on the buttplate. While it was drying he tried one of the screws in a drill gage to see what size drill to use for the holes. He then proceeded to lay out the hole locations with an oddleg caliper and a scriber, and centerpunched them. That done, he gripped the unfinished buttplate by its edges in the drill press vise, using two pieces of thin plywood as yielding jaw inserts.

After drilling the holes, he used a homemade countersink cutter, under the influence of heavy oil and a reduced spindle speed, to produce a beautiful chatterless countersink on each hole.

Dave winked at Sam as he finished champhering the second hole. Some months before, he had made the cutter based on an item he had noted in a book on clockmaking, and had taken much satisfaction in amazing Sam with its chatterfree performance.

"That I should live so long, and then be shown that by my own apprentice!" Sam had laughed. Together, they had then proceeded to make several such cutters* for various purposes, one rainy Saturday afternoon. The wink now recalled the incident completely, for both of them, without need of words.

* see page 53 in this book

176

Chapter 7
SAM GETS IMPATIENT

Dave's phone rang about 9 the next morning at work. It was Nora. "Grampa was awfully sick last night, Dave. He was up half the night, vomiting something terrible. He wouldn't let me call the doctor during the night. I've got an appointment for him for this afternoon. I thought I'd let you know about that. I really don't think he'll be up to having you over tonight, Dave."

"That's okay. I'm sure not very happy to hear about him being so sick. Let's hope the doctor can find out what the trouble is, and give him something to help. If there's anything I can do, let me know."

"Thanks - if there's anything he needs, I'll call you."

"Not just if Sam needs anything, Nora. If you're tied up looking after him, or don't want to leave him alone for long, I'll be happy to do your grocery shopping, or anything else you need done."

"That's... that's very good of you, Dave. Thank you." For a moment it almost seemed as though there was a chink in her armor.

"You'll let me know how he is this evening?"

"I will. I'll let you go for now. Bye."

He went back to his work, disturbed by the phone call. What could be the matter with Sam to make him be sick half the night? Whatever it was, it was not good.

He was in the middle of his supper that evening when Nora called. Sam wanted to see him. Would he come down, but try not to stay too long?

"Okay. I'll be on my way in a few minutes."

She let him in at the front door. "Grampa's downstairs. You'll find he looks quite tired."

He nodded and disappeared into the basement. "Hi, Sam. How you doin' tonight?" He greeted the old man cheerfully, though Sam's appearance did nothing to make him feel cheerful. He looked 5 years older.

"I'm okay. Doctor says I've got an upset stomach - other than that I'm fine. I'm not feeling up to much tonight - couldn't sleep last night. But I wanted to talk to you, Davy. When will that Douglas barrel get here? And the wood for the stock?"

"Well, the barrel could be here in 2 or 3 weeks - maybe a month, what with Customs and all. I ordered it on my credit card on Tuesday, the day after we decided what we wanted. And I had a call back about our fiddleback maple this afternoon. The guy said he can get us some nice wood, but it'll take him about a month to get it in."

"...and the chambering reamers?"

"They'll probably take about a month too. Might be a little sooner."

"That's going to be kinda slow. I was rummagin' around down here today, and I found two odd pieces of plain walnut big enough for a buttstock and forearm. And I phoned Sid Cooper this afternoon. He's got a new 2-groove Springfield barrel and a pair of .30/40 Krag chambering reamers. He'll sell us the barrel and the reamers for $35. Would you be willing to go with a 2-groove military barrel and a plain walnut stock for now? I'll pay for this stuff and we can get on with it right away. We can swap the Douglas barrel for the Springfield barrel when it gets here. What do you say?"

Dave nodded, inwardly thoughtful. It wasn't normal for Sam to be in such a hurry in the development of a project - usually he had the patience of Job. "Sure. How be I go see Sid after work tomorrow, and we can fit the barrel Saturday morning, and maybe make a start on the stock the same day?"

"Okay, that'd suit me fine. We could work on the stock Sunday afternoon, too..."

"If you like. And you can tell me some more about casehardening while we work, if you feel like it."

"I'll feel like it. I'd like to tell you some more tonight, but my throat is kinda raw from being sick last night."

"I can believe it. Look, I've got some things to do at home tonight, particularly if you're going to work me double shifts Saturday and Sunday, so I won't stay longer. Will you be coming upstairs now, or are you going to stay down here for a while?"

"No - go on up and say good night to Nora, and tell her I'll be up in a few minutes."

"Okay. I'll see you Saturday morning... with that 2-groove barrel, and Sid's .30/40 Krag chambering reamers," he grinned, making a circle with his thumb and forefinger.

"Okay - that sounds good, Davy."

Sam's pleasure was evident in his voice and smile, though less so than would have been the case even a week ago.

Chapter 8
A BUSY SATURDAY

Saturday morning Dave Thomas showed up at Sam's basement door with the 2-groove Springfield barrel and the pair of chambering reamers, the latter carefully wrapped in old toweling.

Sam had been feeling better the last day or so, and was nearly his old self. After he'd finished looking at the barrel and the reamers, Dave grinned at him and said "Hey! I gotta tell ya: I lucked onto something else as well."

"What did you get?"

"While I was at Sid's shop, a fella came in with 5 boxes of once-fired .30/40 brass, and a set of reloading dies in .30/40 Krag, wanting to sell them. Sid just pointed him at me, and I bought them, for cheap. I didn't bring the dies down to show you, but they look like they've never been used. So how about that - we're fixed up with brass and a set of loading dies, just like that," he said, snapping his fingers.

"Well, that's first class." Sam said, moving his stool to a spot from which he could comfortably watch Dave work at the lathe. "Now let's get down to some serious business," Sam said, "That Springfield barrel ought to be about 24" long. We were going to put a 26" barrel on the Sharps, but we'll be lucky if this one ends up 23. Will we have to cut a new barrel shank altogether?"

Dave shook his head. "No. Springfield barrel shank specs show the minor diameter of the thread as 0.990". The Sharps Borchardt shank is 0.945"Ø max, so it's all inside the Springfield shank."

He passed Sam a piece of paper with a drawing on it.

"I've been thinking," he went on, "about how to do this, and it should be easy enough. You want me to tell you, or just do it?"

"Go ahead. I'll watch. You know what you're doing." Sam sat back on his stool and watched as Dave slid the barrel into the lathe spindle.

When he had the barrel set up and indicated true, Dave started the lathe and began to modify the Springfield barrel shank to fit the Sharps action. The rough work was quickly done, and the shank brought to 0.945"Ø.

When he was ready to cut the thread, Dave took a special threading toolbit from a drawer in Sam's toolbox. Under Sam's direction, he had filed it from a piece of square drill rod, then hardened, tempered, and finally stoned its cutting edges to a glass-like finish. The final stoning was the secret behind the toolbit's longevity and part of the secret of the flawless thread finish it would produce.

"If you want a nice finish on a broad cut, whether it's a thread, or a radius or whatever, turn slow and use lots of thick cutting oil on the final cuts..." So Sam had taught him long since.

He cut the thread at about 350 rpm until it was nearly done, then dropped the spindle speed to 175 for the final passes, testing the fit of the thread against the Sharps Borchardt action between cuts. Finally, satisfied, he switched off the lathe motor. He broke the sharp crest of the first turn of the thread with a pillar file, and filed off the trailing edge of the final thread where it feathered out into the thread relief groove.

"Now then, what's my barrel OD supposed to be at the breech end?" he asked Sam, referring to the sketch he'd given him earlier.

"1.055 inches."

"Okay. I'm going to give 'er a straight cylindrical form for about 2" ahead of the receiver, and we'll figure out later what kind of a taper we can cut on the rest of the barrel."

The rechambering operation* was next, and here again, Sam's training came to the fore: "If a reamer wants to chatter, or if you don't want it to, just fill the flutes with Crisco." Dave reached for a small screw-top jar of

see note re this at page 126

Crisco after he had unwrapped the chambering reamers.

"You know, I read somewhere that if you wrap a couple of turns of waxed paper around a reamer it'll also cut the tendency to chatter," he said to Sam at one point, when he stopped to wipe accumulated metal cuttings and Crisco from the roughing reamer. He scraped chips and shortening from his index finger onto the rim of the Crisco jar.

"Never tried that. It'd probably work. But I like Crisco," Sam said. The one pound box which Nora kept for him in the refrigerator had a date on it indicating it had been bought 15 years previously. "Lasts a long time, too. You'd only get one use out of a piece of waxed paper." Sam was not one to waste things.

Dave nodded solemnly. "Yeah, ridiculous waste."

They heard the phone once, late in the morning, and Nora called them for lunch just as Dave completed the rechambering job.

"We'll turn the outside taper after lunch," Dave said as he washed up at the laundry sink.

Nora was unusually quiet as she served lunch: thick pea soup and slices of well buttered homemade brown bread. Sam gave thanks, and they ate. There were cinnamon buns for dessert.

"Dave, could you run me down to the grocery store after lunch? I need to get some flour and a couple of other things. Do you mind?"

"No. Happy to. You mind a bit of a delay, Sam?"

"Oh, are you in the middle of something you don't want to leave?" Nora asked.

"Nothing we couldn't leave for lunch, and nothing we can't leave for another half hour, Lass. Off you go. I'll have a nap while you're gone."

"Sam seems better today. He's been enjoying himself this morning," Dave said to her as he started the Olds. He was totally unprepared for the flood of tears that followed.

"What's the matter, Nora?" he asked, reaching for her hand.

"I'll tell you after... please ... let's go."

"Okay," he said quietly, backing out of the driveway. By the time they reached the supermarket she had regained her composure.

"I'll just be a few minutes."

"I'll come in with you, and give you a hand." He went with her, glad to be able to do something for her. Her shopping did not take long, and they were soon back in the car.

"Dave, can we stop somewhere quiet to talk before we go home?"

"Sure. Would you like to sit on a bench along the sea wall?"

"That'd be fine."

The walk along the sea wall was nearly deserted. They sat down on a bench a short distance along the way. He waited for her to open the conversation.

"Did Grampa tell you what was wrong with him?"

"He said the doctor said he's got a bit of an upset stomach."

"That's what he told me, too. That phone call this morning... that was the doctor. It's not an upset stomach, Dave."

"I thought it might not be."

"He said Grampa has cancer of the esophagus. There's nothing they can do for him - it's too far advanced, and even if it wasn't, at his age he probably wouldn't survive an operation to remove the affected part."

Dave sat silent for a few seconds. "Did he say how long he might have?"

"Maybe a month or two - maybe less."

Another silence. "You think he knows all this?"

She nodded, her eyes full of tears. "I'm sure he knows. He doesn't want us upset about it. He's not worried for himself. He...." She stopped, unable to say more.

"Well, if he wants to call it an upset stomach, I guess we should go along with that for as long as we can."

She nodded, wiping her eyes with the heel of her hand. He produced a clean handkerchief and gave it to her. "Now I understand why he's so anxious to see this rifle put together in such a hurry. It's not like him."

After that, neither said anything for several minutes. They sat, seeing but not watching an outbound freighter on English Bay.

Finally she spoke again. "Grampa's the only father I've ever known. You have no idea the things he's done for me, Dave. The least of it is that if it wasn't for his and Granny's determination to get me the best treatment they could, no matter what it cost them in time, trouble and money, I'd have spent my life in a wheelchair."

"I know. He told me that one day. He didn't take so much of the credit, but I figured that was about how it was."

"It hurts so much when you know you have to give up somebody you love, and you can't even tell them how you feel."

"You'll have time to tell him, Nora. He'll see to that, 'cause he knows you need to. I think you'll find he knows how to say good-bye so it doesn't hurt so bad."

She cried again then. He put his arm around her shoulder and pulled her close; he let her cry for a minute or two, to get it out of her system, then told her gently, "I lost my mom to cancer when I was 18. I know something of how you feel, and if I can be a help to you in any way, over the next few weeks, I'd like to be."

"Thank you. Thank you very much, Dave." The shift in the conversation helped her to get control of herself again.

"I've got two weeks holidays coming to me. I can take them just about any time I want, one day at a time or all in a block, so I can be available just about any time, if you need any help."

"Oh Dave, you won't want to use up your holidays for..."

"For you and your grandfather?" He smiled at her. "Now, how about you dry your eyes and blow your nose, and we'll head home?"

She nodded. "Yes, we'd better go. Thanks for... for letting me talk to you." She gave his hand a squeeze and stood up. He'd have walked 50 miles for such a gesture from her.

He and Sam finished the barreling job that afternoon. Dave set the 2-groove Springfield barrel up with the breech end in the Sebastian's 4-jaw chuck, and the muzzle on a live center in the tailstock, and with the aid of the taper turning attachment, contoured the outside of the barrel to suit his own and Sam's ideas of what could best be made of the ex-military barrel. Afterward, they re-crowned the muzzle.

Dave agreed to glass bead blast the barrel at a shop he knew of, and on an off chance, phoned and found the owner at work in spite of the fact that it was a Saturday. Could he come over and bead blast a rifle barrel? Yes, by all means.

So ended the first day of his and Sam's last days together. Tomorrow they would start on the stock.

Chapter 9
STOCK WORK, AND MORE ON CASEHARDENING

"I'm kinda disappointed about that barrel, Sam. The bead blaster was loaded with coarser beads than they usually use, and it didn't make a real nice job of it."

Dave had arrived at Sam's house Sunday morning a few minutes after 9, and they were soon in the shop. He had brought along the barrel, which they were now examining.

"Well, it's not exactly the handsomest job, but it'll do for now, Davy. When you get the Douglas barrel, you can fit it in place of this one and you can bead blast it when you know what's in the machine, or give it a proper polish job, if you prefer. Don't worry about this one - it's only temporary."

Dave nodded. They were both somewhat disappointed, but they would not dwell upon what they couldn't remedy in the time that remained to them.

"Let's get on with laying out that stock," Sam suggested. He began screwing the barrel hand tight into the Sharps Borchardt action as he spoke.

"Okay. That's your specialty." Sam had an eye for line and form which Dave simply did not possess. Sam had set the pieces of walnut out on the bench the previous evening. A few days before, Dave had found a photograph of a Winchester Hi-Wall hunting rifle of very handsome lines in the 1953 Gun Digest*, and had shown it to Sam.

*(page 85; or see page 309 in the Gun Digest Treasury, 4th edition)

Sam now laid the barreled action on the walnut buttstock blank, and with one eye on the photograph and one on the components before him on the bench, drew the profile of the Hi-Wall's buttstock freehand onto the wood, altering it as he did so to suit the Sharps Borchardt's actions' lines and angles, which differ somewhat from those of the Winchester Hi-Wall. He also drew in the axis of the hole that they must drill from buttplate to pistol grip to accommodate the long draw bolt that would anchor the buttstock to the rifle. Dave watched, and shook his head in wonderment: "A veritable Leonardo Da Vinci!"

Sam chuckled that old familiar chuckle, and reached for a bow saw. "Leo DeeVee never had a bandsaw like this one!" Sam always referred to his bow saw as a bandsaw, for it would do almost anything a bandsaw would do, ".... and you don't need electricity to run it, and you're not likely to saw your fingers off in it either." Dave had made the saw as a Christmas present for Sam three years ago. They took turns on the saw until they had the stock blank sawn to Sam's freehand-drawn profile.

"Now then, Davy, you trace that profile onto a piece of cardboard, so when the time comes to make your fiddleback maple stock, you'll have the pattern all ready."

It struck him, as he was carrying out this last instruction, that Sam was deliberately seeing to it that he would leave for Dave a copy of the stock profile, because the day was soon coming when he would no longer be available to provide this sort of help. Suddenly, the sadness he had been holding at bay washed over him like a wave, but he forced it aside, refusing to let it interfere with whatever time they might yet have together.

They drilled the stock bolt hole, and worked on the stock the rest of the day. By late afternoon they had the buttstock almost done. Sam dug out a plastic buttplate, so the ham-

mered steel one could be left till it came time to fit it to the fiddleback maple stock later.

As they worked, Dave had got Sam back onto the topic of casehardening: "... you feel like telling me any more about casehardening, while we're working on this here stock?"

"That's not a bad idea. Where did I get to?"

"Well, you told me how they do it in industry. You said we could make our own pack casehardening compound from charcoal, and you said something about activators. You were going to give me the recipe for your special casehardening mixture, as I recall. That's about where we left off."

"Okay. We'll get to the recipe later. There are some general things you should realize about pack casehardening. For one thing, even the experts are still learning how to do it right. By that I mean it's not as predictable as say a molten salt bath operation, where you know exactly what you've got in the bath, how hot the bath is, and how deep the case will be after so many minutes or hours of carburization."

"But you can do an effective job of casehardening using a carbon pack?" Dave asked.

"Certainly. For most things you'll likely use it for, pack casehardening is fine. You might be making a tool that is subject to knocks an' bumps an' dings an' wear - like that floating arm knurling tool we made last Fall, or a depth gage - and you might want to caseharden it to make it more durable; pack casehardening is fine. You might be making something where a piece of hardened high carbon tool steel is indicated, but you don't have anything suitable at hand. So you make it from hot rolled mild steel, and caseharden it. If you play your cards right, you can end up with a part that is not only tougher, but also its surface may be harder than hardened tool steel would be.

"If you look at something you're makin' and ask yourself, 'What parts in this item would be better if they were surface hardened, rather than just made from soft mild steel?', you'll soon begin to see where to apply the process, and where not to bother.

"Maybe you're making a toolmaker's clamp or a little vise, or something else that's subject to clamping pressures. Or say a set of filing buttons you expected to use several times - assuming they weren't round - otherwise you'd likely make them from drill rod. Or maybe you're makin' a part that will be subject to a lot of wear, or a special hammer that's going to take a lot of shock... those are places were you'd probably want to think about pack casehardening, and it'll serve your purposes quite adequately. You'll have some failures, but you'll learn from them. And if the alternative is to make a part and leave it soft, you'll be back making replacement parts before long, or repairing damage done because something that should've been surface hardened wasn't."

"So don't expect miracles, but don't shy away from doin' it, if it looks like a good idea, either?" Dave said.

"Exactly. Now then, having divested myself of that, let's get on with telling you the HOW of the process."

"Okay."

"The basic approach is to bury the job in the pack casehardening compound, in a steel box. You put a lid on the box, seal the box shut, and heat the whole thing up to carburize the surface layer of the work. Then you dump it into a quenching tank to harden it."

"That's a lotta ground to cover in one session, isn't it? How about telling me how to get a job ready for casehardening?"

Sam settled himself on his stool. "Well, there's not a lot to that side of the matter. Even if you're not after "colors" in a casehardening job, the work must be thoroughly cleaned and degreased - that's the first requirement. And if you do want to end up with a nice play of colors, the job needs to be nicely polished. The better the polish, the better the colors. In that respect, it's like gun blueing: 90% of a real good blue job is in the polishing."

"Okay. What about the box? - you said the work is packed in a steel box," Dave prompted.

"That's right. Actually, in the reducing atmosphere of a charcoal fired furnace, a cast iron box would last a lot longer, but making cast iron boxes isn't a practical proposition for someone foolin' around in his back yard."

Sam paused for a minute, and then went on: "Come to think of it, there are a couple of other sorts of containers a fella could use. One would be a cast iron "Dutch Oven" - if he had one handy that was the right size, and nobody to complain about what he was doin' with it. I wouldn't want to eat out of it afterwards - one of the ingredients· of my pack casehardening mixture is barium carbonate, which is kinda poisonous. The other thing is some sort of a clay crucible."

"Like a flower pot?"

Old Sam's face lit up with a smile, and he waved a gnarled forefinger at Dave. "Now that's about the best idea I've heard all day! A flower pot ought to work like a charm! You'd have to close the hole in the bottom with something - a scrap of a busted pot, say. Then in goes the casehardening mixture, then the work, and more casehardening mixture, heaped up so the pot is overflowing, and finally an old dinner plate, or a big saucer, upside down. Then you'd turn the whole thing over, and pour some dry sand around the exposed edge of the dinner plate for a seal. Plus some more sand on top of the pot, to close off the hole.

"I'd say you might want to try doin' this without a workpiece first time around, 'cause if the pot were to break, due to the heating, your job would be exposed, and it'd scale pretty badly, and real quick, at the temperatures you'd be running."

"Well, it'd sure be worth a try - a clay flower pot would be about the easiest thing to get hold of that ever was. But what about the steel box idea?"

"You can make a box out of mild steel sheet that will do fine; it just won't last forever. That bein' so, it doesn't need to be a high class item. It does need to be big enough to hold the work and then some, and it needs a lid - just a separate flat sheet of steel cut to fit inside the top of the box."

"Okay. Tell me some more."

"You want to make it bigger than the job by maybe 1-1/2 to 2" in all directions. Get yourself some scraps of say 1/8" steel sheet from a steel fabricating shop - they'll have it laying around the shear. Pay the guy a couple of bucks extra to shear the stuff to size for you - two ends, two sides, a top and a bottom. Take 'em home, and weld 'em up into a box with your oxyacetylene torch. The welds don't have to be continuous, like you were trying to make a watertight box - just a short weld in 3 or 4 places along each seam is fine. If you had an electric arc welder, you could tack it together with that. Anyway, make up an open topped box, and if there's any warpage, such that the lid won't go in - remember, it's going to lie directly on top of the charcoal you're going to fill the box with, and you'll seal around the joint between the lid and the walls with fire clay - if it don't fit, file or grind it till it does."

"What about handles?"

"That's something you better do some thinking about. I used to pull my boxes out of the furnace with a big pair of tongs a blacksmith knocked up for me, but I don't know what became of them. You want to make some kind of an arrangement so you can reach into the furnace and connect a good handle to your box, and clamp it on there solid in a hurry. Remember, that box is going to come out of the furnace glowing like an orange, 1600°F thru to the center - you ain't gonna pull it outta there with a pair of oven mitts! And you gotta be able to get the lid off, and maybe tip the box upside down, without it fallin' off your handle."

Dave nodded, smiling at Sam's comment about the oven mitts. "Okay. As far as getting the lid off, a guy could weld a little tab of sheet steel to the middle of the lid - somethin' he could grab with a pair of pliers. I'll do some thinkin' on the matter of a handle. What's next?"

Sam cleared his throat and swallowed. "When you want to use the box, run a fillet of fire-clay paste along the seams on the inside, so it'll be just about, but not quite, air tight. Put about an inch or more of charcoal in the bottom of the box. Then set your workpiece in on top of that, and bury it completely by adding more charcoal. The main thing is that the work must remain completely buried in the charcoal in the box while it's in the furnace. If part of it is exposed, that part won't absorb carbon - which is to say, it won't get case hardened."

"What if you did want to caseharden some areas and leave other areas on the same piece of steel un-casehardened. Can you do that?"

"That sometimes comes up. You might want to machine the bore of a steel gear after case-

hardening it. Or you might want to make a gear and a shaft in one piece, and then machine a spline on the end of the shaft after casehardening. There's two or three ways to do it. What you have to do is mask off the areas you don't want to carburize. One way is to paint the area with fire clay paste well thinned with water. Another way is to copper plate the area, but that's not a practical proposition for a fella tinkerin' around in his basement. The fireclay wash is easy to do - just warm the job up a little so it'll dry faster, and paint it on. Just how certain a cure it is, I don't know."

"You said there are three ways..."

"Yeah. The third way is to buy some goup made and sold for the purpose. One trade name is 'No-Carb'. You just paint it on, and it prevents carburization. There's another way, too: You can leave the job oversize where you don't want to caseharden, and when the work comes out of the furnace, let it cool slowly, then machine the oversized area down to spec, to get rid of the carburized layer. Then you re-heat the job and quench it to get the hardness you want in the case, the case now being only where you want it."

"Okay. When you bury the job in the charcoal, do you tamp the charcoal in tight, or just throw it in loose?"

Sam shrugged. "Just pack it in good and snug; shake the box around, to settle it good, while you're fillin' the box. Like I said, you don't want the charcoal to settle and expose the job during the heating - that's important. If you're doing several parts at once, in the same box, separate them from each other by an inch or so in every direction as you fill the box."

"Then you put the lid on?"

"No. When the work is covered, put in about another inch or so of charcoal, or enough to fill the box up to a point where, when you set the lid in place on top of the charcoal, there's about 1/4" of freeboard left. You want to be sure to have plenty of charcoal on top of the job. Then seal up the lid/box seam from the outside with fireclay paste, so the box is almost air tight, and stick it in the furnace. Come to think of it, I guess you could use sand as a seal on top of the lid, as it'd give a good enough seal, but not so tight it couldn't let out any excess gases that might be given off."

They were interrupted by Nora on the intercom, "Grampa, will you and Dave come up for supper in a couple of minutes, please?"

Dave looked at his watch. "Oh, great - I've stayed right into your supper time."

"Well, there's nobody more welcome at my table than you. We'll be right up, Nora."

"Thanks, but you don't have to plan and cook for uninvited guests. Nora does."

Nora had, too. There was stew and dumplings that would have made a king drool, and lemon meringue pie, after which she told them they had one hour left to work in the shop and then it would be curtains for the weekend.

"I think we're about done with the buttstock, Sam, aside from final fitting and sanding."

"Why don't you leave that for tonight, Davy? I'm about talked out, and it's getting on. We could have a go at the forearm tomorrow night if you feel like coming down..?"

"Sure. It is getting late. Let me clean up this mess a little, Sam, and then I'll let you punch out for the rest of the weekend."

Dave set the buttstock at the back of the workbench, and began to sweep up the shavings and sawdust. "We did pretty fair today, eh? We can whittle the forearm out tomorrow evening, and sand both pieces after that."

Sam nodded his approval. "It's gonna be all right, isn't it?" he said, running a gnarled hand over the buttstock. The old man's pleasure at the day's progress made the effort worthwhile for Dave.

Chapter 10
AN OPAL PENDANT FOR NORA

They got the forearm virtually done the next evening. Sam had been busy in the shop most of the day, working on the items of opal jewellery for the chap from Richmond, and he showed these to Dave and Nora when she brought a tray of hot chocolate and fresh biscuits down to the shop about 9 pm.

"I'll finish them all up by Wednesday," he chuckled, pleased with his day's efforts.

When Nora went upstairs, he dug another piece of opal from his shirt pocket, and showed it to Dave. "I also made a pendant for Nora. Have a look at this." He passed over a beautiful opal teardrop, half the size of a peach stone, with a finding already cemented on the top for a chain. "Say, could you pick up a gold chain for me tomorrow down town?"

"Sure. Any special instructions?"

"No. Just get a really nice one. Take this with you if you like, so you can get a chain that suits." Sam handed him the opal.

"Okay. I'll take it into Birks tomorrow at noon."

"Good idea. Here's some money; if it's any more than that, I'll fix you up with the difference tomorrow night. Get a nice one, and don't concern yourself with the price - get the nicest one they have."

"Okay, I understand what you want." He folded the bills and put them in his pocket. "I was thinking I'd like to put a quarter-rib on this rifle, and mount the scope on that, like a Ruger #1. But that's gonna take time to make. How 'bout for now we just sweat a couple of blocks of steel onto this barrel, mill 'em off level with the bore, and screw a set of Weaver bases onto 'em?"

Sam nodded. "Good idea. Let's do that tomorrow night. Bring your Leupold scope over and we'll see how it's going to come together."

"Okay, I'll do that."

They mounted the Leupold the following evening, and by the time Nora called them up for their snack, they had their .30/40 Sharps Borchardt largely put together. It didn't look too bad, in spite of its rough sanded stock - a diamond in the rough, Sam said, pointing out also that the angle of the bottom of the pistol grip should be altered to look right. Dave could see that this was so, when it was pointed out.

Sam was pleased with the gold chain Dave had picked out for the fire opal pendant. While the three of them were eating their snack, he slid a small box across the table to his beloved granddaughter.

When she opened it, the opal caught the light and gave forth a breathtaking play of pale blues, pinks and greens. Her look of surprised delight brought that familiar happy chuckle to Sam's voice and his wrinkled old face broke into a smile. Nora got up and came round the table. "Thank you, Grampa - it's beautiful." She put her arms around his neck and kissed him. "My Grampa!" she whispered, her eyes filling with tears.

"Put it on, Lass."

Dave did not mention the furor it had caused in Birks earlier that day, nor the several thousand dollars he had been quoted as to its probable value.

Chapter 11
CHARCOAL & FIREBRICKS

"You said you'd show me how to make a furnace. Do you feel like doing that tonight?" Dave asked the following evening, as he began wiping down the bead-blasted barrel with rubbing alcohol. They had agreed to use a cold "touch up blue" solution on the 2-groove Springfield barrel for now, and to do a proper slow rust blue on the Douglas barrel, its quarter rib, and the Sharps Borchardt action later. Dave was degreasing the barrel in preparation for applying the cold blue solution. Sam sat on his stool, watching.

"Sure. There's a couple of ways to go. One way is to make a charcoal fired furnace like the Egyptians used 3500 years ago. The other way is to buy an electric furnace of some sort. How about I tell you about the charcoal fired type tonight?"

"Sounds good."

"Okay. What to do is to get yourself some insulating firebricks, and make a knock-down furnace out of them in your back yard. You'll need a slab of firebrick about a foot square for a cover for the top of the furnace - something you can lift on and off. And a source of air - you gotta blow it to get the temperature up where you need it. Your vacuum cleaner would work fine for that, but you'll want to make some sort of a 'gate valve' so you can regulate the amount of air goin' to the furnace, and spill the excess. Bottom chamber is for the air. Fuel goes on the hearth. Work goes on the firebrick 'splits' above the fuel. The hearth is the only tricky part."

"How so?"

"You need to make a wooden mold - four 1 x 2's tacked to a piece of plywood is all you need - and then mix up some castable refractory clay, and pour it into the mold. Then lay the edge of a board - what a concrete worker would call a screed - across the top of the mold, and saw it back and forth as you take it across the clay to level it out. You want to avoid playing about with the clay once you get it leveled off in the mold. If you start patting and rubbing at it with a trowel, to make it all nice and smooth, you'll spoil it."

"Why is that?"

"Because you'll bring the water to the surface, and you don't want that. A slightly rough screeded surface is what you want."

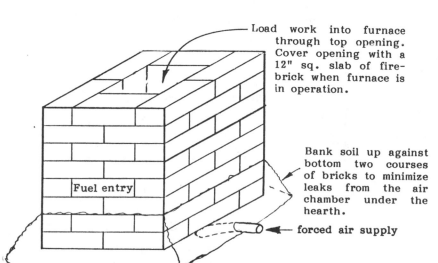

Load work into furnace through top opening. Cover opening with a 12" sq. slab of firebrick when furnace is in operation.

Bank soil up against bottom two courses of bricks to minimize leaks from the air chamber under the hearth.

← forced air supply

Furnace requires 48 insulating fire bricks to construct. Exterior furnace dimensions: 18" square x 20" high; interior is 9" square.

Fuel entry brick should be removable.

Slab for furnace top should be provided with a handle for easy removal and repositioning - suggest stainless steel strapping for this.

See next page for interior details.

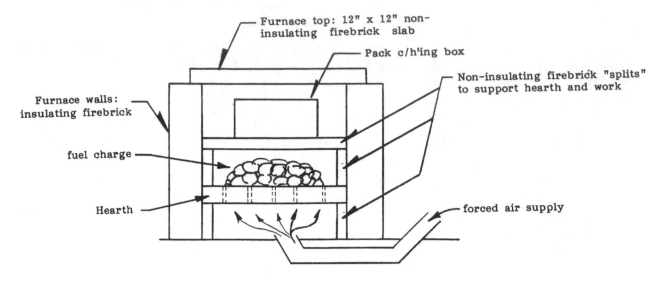

Furnace top: 12" x 12" non-insulating firebrick slab

Pack c/h'ing box

Non-insulating firebrick "splits" to support hearth and work

Furnace walls: insulating firebrick

fuel charge

Hearth

forced air supply

"Okay. What else?"

"The hearth's gotta have several holes in it to let the air supply come up through it to blow the fire. You can push some little sticks - oh, say about the size of a pencil - into the clay while it's wet, and twist 'em out, or drill 'em out, when you're done curing the hearth."

"How do you do that?"

"Let the clay dry in the mold for about 24 hours. Then take the hearth out of the mold and put it in your kitchen stove oven, cold, and turn the oven on to about 250°F, to drive out the water. After 2 or 3 hours, or when it quits steaming, start raising the oven temperature by about 20° every 10 minutes until the oven is up to say 500°, then turn the oven off, and let it cool down. When the oven is cool, the hearth's done, and ready to use."

"And once you've made the hearth, it should last for quite a while?"

"So long as you look after it. The whole furnace can be set up or dismantled anytime you want - it's just loose bricks. You take it apart, and when you get down to the hearth casting, you lift it out and store it for the next time you want to caseharden something."

"Could you make a hearth out of steel rods welded together? Say some re-bar?"

"I don't know - I suppose it would work. No reason not to try it."

"You were mentioning firebrick splits a minute ago, and you spoke of insulating firebrick, and non-insulating firebricks. What's all that mean?"

"A firebrick 'split' is half the thickness of a normal size brick. It'd be say 9 x 4-1/2 x 1-1/4" thick, instead of 2-1/2" thick. Insulating firebrick is made by mixing sawdust with the wet clay. When the bricks are fired, in manufacture, the sawdust burns out, leaving the brick full of voids, and much less inclined to

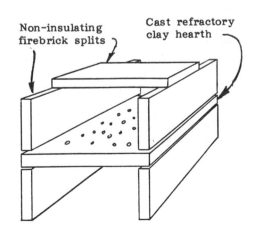

Non-insulating firebrick splits

Cast refractory clay hearth

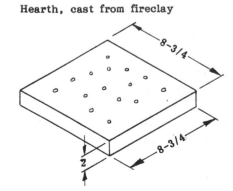

Hearth, cast from fireclay

8-3/4

8-3/4

2

conduct heat. Non-insulating firebricks are heavier, and have no voids, so they carry off the heat much faster. They're also a lot stronger than the insulating type, which will not stand much banging about - insulating firebricks are actually pretty fragile."

"Okay. And what about the actual heating, and quenching and all that?"

"How about I tell you about that tomorrow?" Sam chuckled. "And I'll tell you how to make and operate your quenching tank so you get all those colors you see on a good color case-hardening job."

"Okay. I better give this barrel another pass with the cold blue - it's not very dark yet." He held the barrel to the light so Sam could see it better.

"Yup. 'Nuther pass'll do 'er good. Then it's just about ready, except for sanding and finishing the stock." Sam seemed pleased at the progress they had made since Saturday.

Dave agreed to come again the following evening.

Chapter 12
TO THE HOSPITAL

The phone rang 3 times before Dave woke up sufficiently to realize what the noise was. He rolled over and groped for the phone beside his bed. "Hello?"

"Dave - " It was Nora. He came fully awake. "I'm sorry to phone you at this hour, but it's Grampa - he got up about half an hour ago to go to the bathroom, and he fell in the hallway."

"Did he break anything?" Dave looked at the clock - it was a little after 3.

"No. He's okay that way, but he's coughing up quite a bit of blood. He doesn't want me to call the doctor at this hour of the night."

"Do you think he'd let me take him to the emergency ward? They could have somebody look at him there."

"I think that's a good idea - maybe he would..."

"I'll be right down, and we'll zip him over to the hospital."

He dressed, grabbed a couple of blankets and his car keys, and was out the door. The streets were deserted and he took full advantage of it.

Nora came to the door in a dressing gown. "Thanks for coming so quickly! Come in - Grampa's in the kitchen." She was whispering, more from anxiety than anything else. She looked tired and worried.

Sam was seated at the kitchen table, a cup of warm water in front of him.

"Hi, Sam. How you feeling?" Dave squatted beside his old pal's chair.

"Lousy. Can't sleep. Can't get to the bathroom in my own house without falling down. Can't keep my supper down."

"Did you hurt yourself when you fell?"

"I don't think so. I caught myself on the wall before I hit. Guess I gave Nora quite a scare." He said ruefully.

"That's all right, Grampa." She patted his white hair lovingly.

"Nora says you're coughing up some blood. That right?"

Sam nodded, frowning. He wiped his hand over his face in some distress.

"Would you let us take you over to the hospital and let somebody in the Emergency Ward have a peek at you?"

Sam thought about that for a minute. "You think that's a good idea?"

"Yeah, I do. If you like, we'll take you over there, and stay with you till they see how you are. And when they say you can come home, we'll bring you back here."

"Well, maybe that would be a good idea." There was almost a note of relief in Sam's voice.

"Good. Do you want to get dressed, or go in your pyjamas?"

"Oh, no, I'll get dressed."

"You like a hand to do that?"

"No. I can get dressed on my own, thanks." He wasn't testy about it - it was just a statement of fact.

187

"I'll lay out something for you to wear, Grampa," Nora gave him a pat on the shoulder and left the room. By the time Sam was ready to go, Nora was dressed and had rounded up Sam's wallet and the house keys.

Dave helped him into the front seat of the car, and spread a blanket over his knees. The porch light made the night's blackness seem all the more intense beyond its small circle.

When they pulled up at the Emergency entrance, the hospital staff largely took over. Dave and Nora sat out two and a half hours, most of it in the Emergency Ward waiting area, part of it at Sam's bedside. Finally, the doctor on duty came to the waiting room and spoke to them.

"I'd like to keep Mr. Carter in the hospital at least overnight. He's in no immediate danger, but he is hemorrhaging somewhat in his esophagus. We want to stop that if we can. He's sleeping at the moment. You might as well go home for now. We'll call you if there's any change."

"All right. I'll leave him a note." She borrowed pen and paper from the desk and after a moment's hesitation began to write. When she was done, she handed it to Dave to read. "Is that all right?"

"Grampa:

You were sleeping when we left. The doctor said they want to keep you in overnight for observation. He suggested we go home for now. You can call me, or have a nurse call me, any time you want. I'll come back during visiting hours to see you. Dave will come later, when he gets off work.

Your loving Nora."

"That's good... maybe put your phone number on it, too."

She amended the note, and asked a nurse to give it to him when he woke up.

"...And you will call me if there's any reason to?"

"Immediately, Love. Off you go now, and get some sleep yourself. You've probably been up half the night."

They left, and walked to the car. The streets were still empty as they drove home - most people were just getting out of bed.

"Will you come in and let me make you some breakfast, Dave?"

"No, thanks, Nora. I'm fine. I'll drop you at home for now, and get off to work. If you like, take a cab over to the hospital this afternoon - I'll be along to see Sam after work, and I'll take you home then. That suit you?" She drove, but she did not like to, and Sam's 1948 DeSoto was cantankerous, to say the least. Sam knew all its tricks, and Dave most of them, but Nora's relationship with the old green car was more like an armed truce than anything else.

"That'd be nice, Dave. And thanks so much for your help this morning."

"Ah - that's nothing at all. I'm glad you called me."

He skipped breakfast, showered, shaved, and caught his usual bus to work. The morning dragged, and he phoned Nora at noon to ask how Sam was.

Oscar Matlin answered the phone.

Chapter 13
NEW DEVELOPMENTS

"Oscar? Dave Thomas. Is Nora there?"

"No, she's out. I talked to her earlier this morning. She's fine. I came over about an hour ago to operate Sam's rig. She and Sam were out so I let myself in."

Dave too knew where the spare key was kept, under a rock near the basement door at the back of the house. Oscar Matlin often used Sam's ham radio installation, which occupied a room in the attic.

"Any word on how Sam is?"

"You'll have to ask Nora about Sam when you see her."

Dave's jaw clenched. Trust Oscar to pretend that he, Dave, shouldn't be told "confidential" family matters!

"Oscar, I drove Sam and Nora to the hospital about 3 o'clock this morning."

"Oh - I didn't know you knew he was in the hospital. Nora's there right now."

"Fine.   I'll   talk   to   her   later."   He paused..."Look, I'm sorry if I bit your head off just now.   I'm just real worried about Sam."

"So am I - say, did you know Sam was making a Sharps Borchardt rifle? It's downstairs on the work bench. .... You still there, Dave?"

"Yes.   And yes, I knew Sam was making a rifle on a Borchardt action. I've been helping him with it."

"Well, you'd better bush the firing pin on that action. Otherwise you're likely to get pierced primers. You'll have to open out the face of the breech block and fit a plate with a smaller hole in it, and make a new firing pin to suit."

"We know about that. Anyway, Oscar, I'll let you go for now - I gotta get back to work."

Good old know-it-all Oscar! He couldn't cut a paper doll out of a piece of galvanized sheet metal if the outline was painted on it with a 1" brush, and here he is telling two machinists about bushing a firing pin hole! What next!?

Nora called him late in the afternoon, and asked if he would be coming to the hospital after work as he had planned? Yes, he'd be there about 5. How was Sam?

"He's resting quietly. They've moved him to a regular ward - it's Room 212, on 3 East. The doctor says he can come home tomorrow."

"That's good news, Nora. I'll see you after work. 212, 3 East."

Sam was very quiet. He looked tired and old. Nora herself didn't look her best - she hadn't done her hair or make-up at all that day, and apologized for her appearance.

"You look fine to me. Have either of you had dinner yet?"

Sam had had a little soup for supper, and the same for lunch. Nora had had a sandwich at noon. She was about ready to go home, but wanted to stay with her Grandfather if he wanted her to.

"No. You go on home and get some supper, and some sleep. I'll be fine. Thanks for coming this afternoon to keep me company."

"Would you like me to come over later this evening, before the end of visiting hours?" Dave asked.

"No thanks, Davy. I'm going to nod off pretty soon, I think. I'm quite happy here for the night. Off you go."

Dave asked Nora if she would let him take her to dinner, but she refused, again citing her appearance.

"Nora, you look nicer, to me, just the way you are, than any other girl I know. Come on - let's go get something to eat."

She dug out her lipstick, and there was a real warmth in her smile of acquiescence. Their conversation over dinner lasted much longer than Dave would have predicted, and took a turn which he would never have predicted. When they finally said good night, they were on a new and entirely different footing with each other.

Chapter 14
THE FURNACE AND THE QUENCH

Dave did not see Sam the next night. Nora called about supper time to say she had brought Sam home from the hospital that morning; he was feeling better, but he was not up to even a short visit. By the following evening, Sam was feeling well enough that he asked Dave to come down to visit. Nora warned him to say nothing about Sam's fall and the hospital episode - he seemed to want to treat it as if it hadn't even happened.

"I thought I'd tell you some more about the furnace, and how to rig up a quench tank for color case hardening," Sam said, once they got settled in the basement. Dave noticed Nora had put a portable electric heater in the middle of the shop floor where it'd do Sam the most good.

"I was telling you about how to make a charcoal fueled furnace the other night. With a furnace like that, you're going to have to learn to judge the temperature in it by eye."

"Is that hard to do?"

"We did it years ago, just by experience and 'feel'. Just how close we were to what we thought we had, for temperatures, I'm not sure, but we casehardened lots of stuff," Sam said with a smile.

"I'm not sure whether you told me before or not, Sam, but what kind of temperatures are we talking about?"

"For carburizing, 1650 - 1700°F is the area to shoot for, most of the time," Sam replied. "You don't want to go much above 1700° - if you hit 1800°F, which is near a yellow color, you'll burn the steel, and ruin it completely. I'll tell you about the temperature routines tonight, if you like."

"Okay - I'm all ears."

"You can buy temperature indicating pellets,* and I think a fella could train himself pretty well with them. Say you set three of them on top of the pack box, when you put it in the furnace, and then watch them as you run the heat up. The first one would be a 1600°F pellet, say, and the next one would be 1650°, and the third one'd be for say 1750. When the first pellet melts, you know you're getting near 1650. When the 1650 pellet melts, you want to notice the color of the interior of the furnace very closely, and then try to keep it at that same color from then on. If the 3rd pellet melts, note the color at that point also, and then cut back on the air supply right away, so's to drop the temperature back - you don't want to exceed 1750°F."

*See Appendix for source

"Could you drop the temperature back, stick another 1650° pellet on the pack box, and then bring the heat back up until it melts?"

"Yes. That's the idea."

"So like you said, a guy could train himself as much as he needed to, to be able to recognize what the furnace interior looks like at 1650°F?"

"That's right. But there is one flaw in the idea, and it's this: the interior of the furnace will look quite different on a bright sunny day than on a dull cloudy day, so you might go through a lotta pellets." Sam smiled.

"I said the other night that the other way to go is with an electric furnace. That's a little more civilized all around, particularly since we're talkin' about "soaking the work at temperature", as they say, for maybe 2 to 5 hours, or more, depending on the depth of case you want. If you bought a small electric heat treating furnace*, you'd be able to dial up the temperature you wanted, and the furnace controls would keep it there for you. I think if I was setting myself up to do case-hardening in my basement today, that's the way I'd jump. We didn't have that kind of equipment available years ago."

*Brownell's, Inc. sells a small 110/220V electric furnace priced about $350 that looks like it would be suitable. It goes up to 2000°F and has a heating chamber big enough to accommodate the sort of small parts most of us would want to do. A potter's kiln would also serve, but would cost more.

"Maybe the thing to do would be to try it with the charcoal furnace at first, and then if I see I'm going to do much of it, and don't find judging the temperature very easy, I could get an electric furnace then."

"That makes sense," Sam agreed. "Now, the next thing I was going to tell you about was the quench tank, and quenching the work. What you gotta have is a tank of water, with an air hose connected to the bottom at the side, like this." He passed Dave a sketch he'd made earlier.

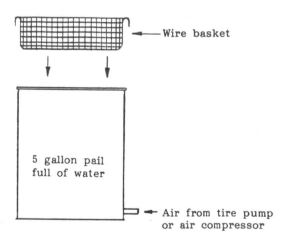

"Saves me talking so much," he explained. "Now, if you want a nice play of colors on the job - and you don't necessarily always want colors - when you dump your job into the quench tank, you want to have a steady stream of air bubbles - lots of bubbles - rising through the water. That helps to get the colors. Don't do no harm to have a touch of oil floating on the surface. And you can add a whiff of table salt to the water also, if you like."

"And you just dump the whole mess into the tank? Seems to me it'd blow right back out of the water at you!"

"Well, you don't throw the whole pot in there!" Sam laughed. "You give the lid a clout to loosen it, and you get it off the box, and then you fish the job out and dump it into the tank by itself. The less time elapses, and the less chance the air has to get at the work, the better."

"Okay, I get you." Dave said.

"If you're not after colors, just forget about the air bubbles. If you want a really hard case, you can use a brine quench, 'stead of plain water - brine is generally a very good quench for mild steel. Another thing: the water in the quench tank should be cold, and it should be "soft" - you don't want to use hard water. Our water is soft right out of the tap, here in Vancouver. But in lots of places, the water is hard - full of minerals. If you lived in such an area, you'd want to collect some rain water, and save it for quenching."

"Now there are several ways you can jump when you take the work out of the furnace to quench it. If the piece is something non-critical, like our buttplate, there's no reason to do any more than just knock the lid off, dig the work out of the charcoal and stuff it into the quench tank, and leave it at that. That'll give you a soft, tough core, and a very hard case, providing the work was hot enough for the carburized layer to harden when it went into the quench tank."

"Which it would be at 1650°F." Dave added.

Sam nodded. "Right. Now, if you quench direct from the carburizing temperature, you are likely to end up with a rather coarse grain structure in the core, due to the length of time the work was soaked at the carburizing temperature. That coarse grain structure doesn't make for maximum toughness. If you want the finished part as tough as possible, you have to double heat treat the part, to develop a finer grain in the core."

"Isn't that getting too sophisticated to do at home?" Dave asked.

"No, not for the sort of parts you're likely to be making. Naturally, I'm not recommending you try making and heat treating something critical, like a new bolt for a rifle. But if you think a particular job demands it, you can do more than just whip it outta the furnace and dump it in a 45 gallon drum full of cold water."

Dave laughed. "Okay. Tell me what to do."

"You take the pack box out of the furnace, and let it cool down slowly with the work still in it. Then dig the work out of the pack box, and repack it in fresh charcoal - and here you don't need any activators, 'cause you're just using the charcoal to prevent scale formation and loss of carbon; then re-heat it to say 1450-1500°F, and quench it again."

"How long would you hold it at that temperature?" Dave asked.

"Well, the work, not just the outside of the pack box, has to get up to temperature. Until the work is up to temperature, the box will look like it has a darker area in the middle - it's a funny thing, but it's almost like you can look right into the box. Once the box is the same color all the way through, the work is up to temperature. Then you can take the box out of the furnace, get the job out of the pack, and into the quench.

"The point is, this brief second heating to a somewhat lower temperature, followed by quenching, will give you a much finer grain structure in the core than you'd have if you'd just quenched the job after the prolonged carburization heating at say 1700°F. The end result will be a core as tough as wet rawhide, and a case as hard as a witch's heart."

"What about tempering that surface layer, Sam? Is that necessary?"

"Well, it's never a bad idea, just to take the stresses out, if nothing else. You can heat the job up to say 350° - 400°F for 20 to 30 minutes, and then quench it again, in water or oil."

"If you want to know more about it, I'll give you a book..." Sam got up and went to a cupboard in another part of the basement, and returned with a book which he handed to Dave.

"*Hatcher's Notebook!*" Dave exclaimed. "I didn't know you had a copy of this! Man, I've always wanted to read this!"

"Well, you take it home and read it. You'll find there's a chapter in there that'll tell you in detail about how they heat treated the Model 1903 Springfield receivers and bolts over the years. You read that chapter in old Hatcher's Notebook two or three times and you'll know enough to do just about anything you'll ever need or want to do in the way of pack casehardening in your back yard. Hatcher will tell you stuff you'll never do, but it'll give you a better understanding of what I've been telling you, if you want to delve into it some more. What I've told you is enough to do just about anything you're likely to need to do, but it never hurts to know more about it. I think I've got a copy of *Metals Handbook* around here too, from about 1948. There's some good gen in there too, and in *Machinery's Handbook* as well. You take a read through them now and then, and you'll pick up some more ideas each time you do, now that you understand the basics."

"On something like our buttplate, how long would you have to keep it in the furnace to carburize it?" Dave asked.

"On a thin part like that, I'd heat it up to 1650°F, and hold it there for maybe an hour. It's thin, so you can't leave it too long, or you'll cook the carbon right through to the middle from both sides, and end up with a buttplate as hard and brittle as a file. That you don't want. The main reason we'd have for casehardening the buttplate, aside from the practical desire to make it more resistant to damage, is to get a nice show of colors on it. Like I told you earlier, the way to get that is to have a mess of air bubbles rising through the quench water when you dump the work in. And some oil on the surface of the water, and some salt in the water."

Sam paused to scratch his ear, and then continued. "That's about all there is to it. From there on, you just gotta get in there and try it. Besides, my throat doesn't feel like much more talking tonight....unless you have some other questions?"

"Not right now I don't, but I'll probably think of some later. Do you want to do some more work on the Sharps?"

"There really isn't much more to do. We should bush the firing pin, but we don't have to do that immediately. It's about ready to test. What do you say we take it out to the rifle range on Saturday?"

"Sounds good to me, if you feel like doin' that. I'll make up some ammo tomorrow night - just moderate loads - and we'll give it a try. I don't suppose we'll set any accuracy records first time out, though - we may have to spend weeks working up a good load, seeing what it likes in the way of forearm pressure, and so on."

Sam nodded. "You want to try some jacketed bullets in it to start with? That'd probably make it easier to get some sort of decent accuracy right off the bat. Cast loads can be pretty finicky sometimes..."

"Well, we can if you want, Sam, but I have some of those pointed Lyman 311365 .30 caliber bullets all cast and sized and lubed... We usually have good luck with that one if the rifle will shoot at all.* We said we were going to make this a cast-bullets-only rifle, so why don't we try it that way first time out?"

"You're right. If it don't shoot all that hot, it won't be the end of the world - we'll at least have had the day out at the range. Maybe after that we can bush the firing pin, and install a set of sling swivel bases."

Dave nodded. "Yeah - we could do that Sunday afternoon. But look, I shouldn't stay any later tonight, Sam. If I don't talk to you before Saturday, I'll see you about 10 o'clock that morning, okay?"

Chapter 15
SAM'S RECIPE

Saturday, so far as the rifle's performance was concerned, was not a red letter day. Groups hovered around the 4" mark at 100 yards, certainly nothing to gladden the heart of a rifleman, and extraction of fired cases was somewhat difficult.

They stopped for lunch at a favorite restaurant on the way home, and over two steaming bowls of clam chowder, Dave reminded Sam that he had not yet told him the recipe for his special casehardening compound.

"Do you feel like another ear full right now?" Sam asked.

"Sure."

"Okay. Like I told you before, what you want to do is make yourself some high grade charcoal. First thing you'll need is a bunch of fruit pits and nut shells. Apricots, peaches, cherries, plums, prunes - anything that has a hard stone - and walnut shells, almond shells, coconut shells - anything like that."

"Getting fruit pits should be easy: next canning season I'll ask a couple of neighbor ladies who put up a lot of fruit to save all their fruit pits for me."

"That'd do it. When you get 'em, spread 'em out on a newspaper to dry for a day or two in the sun. That way, when you go to make them into charcoal, they'll be dry, so they'll heat up faster than if they are all soggy. And if you don't get to them right away, they won't go all moldy on you, or attract bugs.

"In the meantime," Sam continued, "if you can scrounge some, you can also use hardwood, the denser the better: hickory, oak, ash, walnut, elm - that sort of stuff. I've heard that oak is best."

"Okay. I know a cabinet maker who'd have lots of hardwood scraps. Would branches off a hardwood tree be any good?"

"If they're good solid wood, but you don't want new growth. And you want to cut it up into quite small pieces, say smaller than your thumb."

"Okay. Suppose we lump all this stuff together and call it fruit pits." Dave suggested. "What do we do with it?"

"We turn it into charcoal, and we impregnate the charcoal with a couple of 'activators', and then use the charcoal as a pack casehardening compound."

"How do you make charcoal?"

"Just heat the stuff in the absence of air. It's very easy to do on a small scale. You won't need much to caseharden that buttplate - you can make a good supply using a couple of soup cans in your fireplace."

"How do I keep the air out?"

"Just use an open topped can, and put another can over it, upside down. You don't need a sealed air-tight fit - just something that'll drop down loose over it is fine. A canned salmon tin makes a good lid for a soup can. Punch some holes - 3 or 4 maybe - in your lid. Pack a bunch of fruit pits or other stuff

into the soup can, put the lid on it, and set it in the fire. If the lid looks like it wants to come off while the gases are burning off, you can put a little rock on top to hold it on. While the first can is heating, fill the other one, put a lid on it, and have it ready to go into the fire when the first one's done."

"Would it be smart to make up a heavier container, say out of pipe fittings, or 1/8" sheet steel?"

"No. That'd just slow the heating process. The cans are just throw-away units, and being lightweight, they heat up fast."

"What happens when you put 'em in the fire?"

Old Sam's eyes twinkled. "Well, Davy, I suppose I ought to leave you to see for yourself what happens, but I'll tell you. The heat from your fire - and you want to set the can right in among the hot coals - the heat drives off the volatile gases in the fruit pits. In a big commercial charcoal making operation, they would recover those gases and make acetic acid, methyl alcohol, acetone, and tar from them, but we just want to get rid of them.

"The holes in the lid vent the gases. At first you'll see smoke coming out of the holes, then pretty soon the smoke will ignite, and you can tell when the whole business has come to an end, because the flames will stop whooshing out of the can. After that, pull the can out of the fire and leave it to one side for a few minutes to cool with the lid on, and put the next can in the fire."

"And what do I do with the charcoal when I'm ready to take the lid off the can?"

"You want to have something fireproof to dump it out into - a tin pan, or a Pyrex bowl, or whatever. Then fill up the can again, and put it back in the fire. And you'll get quite a surprise when you take the lid off that first can - the fruit pits shrink to about half their original size."

"Because of all the gas that's driven off in the charcoalizing process?"

"I suppose so... anyway, that's what happens."

"How much charcoal do we need?"

"If you can collect two or three gallons of fruit pits, shells, wood and other such like, you'll have more than enough for half a dozen buttplates. You can always use it on something else later."

Dave nodded, thinking. "And when it comes out of the fire, and cools, it's ready to use?"

"Not quite. You have to crush it, and screen out all the fine stuff, which we don't want. Charcoal from say 1/8"Ø up to about say 3/8"Ø size is good, so you want to get yourself a couple of pieces of screen with mesh sizes of about 1/8" and 3/8" or so. Run all your charcoal through both of them. Whatever stays on top of the fine screen is the stuff to use, and anything that doesn't pass the coarse screen should be crushed some more - you can crush it by putting it in a bag of some sort and stepping on it with your foot. Then shake the bag around, and do it again."

"Another thing: this whole business of crushing and screening is best done outside - the fine charcoal dust is just like flour, only it's as black as the inside of a cow - not the sort of thing you want to have all over the shop floor."

Dave nodded, taking in all that Sam had told him. Then he asked, "You said something about 'activators.' What about them?"

"The activators," Sam explained, "are powdered barium carbonate and sodium carbonate. When you heat the activator/charcoal mixture, the carbonate decomposes to a barium- or sodium-oxide molecule plus 2 carbon monoxide molecules. One carbon atom from the 2 carbon monoxide molecules goes into the steel - that's the part we are interested in - and what's left becomes carbon dioxide, which recombines with the barium or sodium oxide molecule, which gives you the carbonate form again."

"I never had much of a feel for chemistry," Dave said, wrinkling his face.

"Well, look at it this way: The activators promote the transfer of carbon from the pack into the steel. It don't matter much just how or why. What does matter is that they make a big difference - without them it might take 8 to 12 hours to get the same depth of case you could get in 2 hours with them."

"Well then, I guess I'll just have to learn to love them." Dave said with a grin. "What do

you do with them - just mix them in with the charcoal when you're packing the work in the pack box?"

"No. You put the barium carbonate and sodium carbonate in a glass bowl, and add enough water to make a sort of "activator soup". Some of the powder won't completely dissolve, so stir it around real good to get it into suspension - then dump the whole thing over the charcoal and let it soak it all up. I'd say about 1/2 pound of carbonates to 2 pounds of charcoal, and use about 80% barium carbonate to 20% sodium carbonate in the first place. Apparently any more than 8% carbonates is wasted, but anything less than 8% is ineffective, and since you can't add more activator while the process is underway, you've gotta put in more than enough to start with."

"Okay. And then what?"

"When the 'soup' is all soaked up by the charcoal, spread the charcoal out on a sheet of paper to dry. When it's thoroughly dry, your homemade pack casehardening compound is ready to use. It's the Bullseye Mixture."

"I bet it is. One last question, Sam: where can I get barium carbonate and sodium carbonate?"

"The sodium carbonate is easy - just buy some washing soda in the super market. Barium carbonate is used in the glazing of pottery, so look in the Yellow Pages under "potters' supplies" and you'll have a source. If you had to get the barium carbonate from a chemical supply house - say if you lived in a small town, and had to order it from a big center - you wouldn't need to buy the high purity grade. The regular 98% purity stuff is all you need."

"Didn't you say the other night that barium carbonate is poisonous?"

"Yes, I did. Don't eat it. You don't want to breathe it either, and you want to wash your hands and face after you've been working with it, but that's about all. There might be some more safety instructions on the package - if there are, see that you read 'em.

"And another thing: since what goes on in the carburizing pack doesn't lead to the loss of the barium carbonate, it's still there when you eventually get around to thinkin' about getting rid of the used compound in the pack box when it's cold, so don't just chuck it in the garden. I suppose worse things get thrown into the sewers, but if it was me, I'd put it in a bottle and when I had a chance, I'd give it to the government people who handle hazardous wastes, and let them look after it."

Chapter 16
TO EVERYTHING THERE IS A SEASON

Sam was sick again the next day. He had asked to be taken to church, as had been his weekly habit for more than 60 years, but he had had to cancel the idea - he was so tired out from the effort of dressing that Nora had simply said, "We will have church at home today", and she and Dave had read several passages to Sam from his worn old Bible, some at his request, some of their own choosing.

The three of them sat quiet then for a while. After a time Sam looked up at the clock on the mantle, and when he spoke, he almost seemed to be talking to himself: "To everything there is a season, and a time for every purpose under heaven - a time to be born, and a time to die, a time to mourn and a time to laugh... Naked came I into this life, and naked I depart..." He looked back to Nora and Dave. "You are my two last remaining loves. I feel that the time has come that we must say good-bye."

"Oh, Grampa..."

"Don't be sad, Nora. No man could ask for more than I have had in this life. We must all die one day. I feel it in my bones that I will not see the sun set this day. I want to thank you both for the way you have helped me this last while..." he paused, breathing slowly. "Nora, I want to talk to Dave for a few minutes. Maybe you could go and make the two of you something for lunch...?" Nora nodded, and left the two men alone.

"Dave, you take good care of Nora. She's going to need someone to lean on, especially over these next few weeks."

"I will, Sam."

The old man held out his hand, and Dave took it. The grip was weak, far from what it had been even two weeks ago. "Just to shake your hand is like a day's work, Davy. Ah,

but we've had some fine times together, haven't we, son?"

"We sure have, Sam." He smiled and squeezed the old man's hand. They sat in silence for a time, then Dave spoke. "I'll never forget you, Sam, as long as I live. You're the finest man I've ever met. I've learned a lot from you... I don't mean just the shop stuff, but your priorities, and the way you treat people - I think that's the *real* Bullseye Mixture."

"So it is, Dave. So it is."

Sam was quiet for a while before he spoke again. "I want you to move the shop up to your house - all of it - the machines, and all the tools. That's been my intention ever since I first saw how interested you were in learning the trade. All through the years I was building up my shop, always in the back of my mind was the hope that before I passed on, I'd find somebody who was worth turning it all over to. There's nobody I'd rather see have my tools than you, Davy. And when your stock blank and that Douglas barrel get here, you fix up that Sharps and make it a rifle we'd both be proud of. I'd like that."

Dave nodded wordlessly, fighting to retain his composure.

Sam was quiet for several minutes before he spoke again. "I want to have a while to talk to Nora, Davy. There's a box downstairs on the workbench - some other things I want you to have. Take it home, and have a look through it. Nora will give you a call later this afternoon. Okay?"

"Okay, Sam. I'll tell Nora you'd like to talk to her, and I'll go home for now."

"Yes. I may not see you again, Davy. Safe journey."

Dave leaned down and pressed his head against the old man's white hair. "And for you, Sam - Safe Journey for you too." He straightened up then, smiled at his old friend, and went to the kitchen, where he wiped his eyes unashamedly. "Sam wants to talk to you, Nora. He said there's a box for me downstairs on the workbench..."

"Yes, I helped Grampa pack that last night. You get it, and go on home for now. Thanks for being here this morning - I really appreciated the passages you read. Oh, here - I made you a sandwich. I'll wrap it and you can take it with you. I feel like I'm trying to shoo you out of the house, but I'm not."

"No, no. I know. Thanks - you didn't need to do that..."

"That's ok - it helps to have something to do. I'll call you later this afternoon, or immediately if Grampa shows any sign of a change."

"Okay, Nora. You go on in and talk to him, and I'll go now."

He didn't have much heart for looking in the box that afternoon, but he did open it. In the top was Sam's "Kowa Six" camera, an equally fine light meter, and 2 extra lenses. It was a very good camera, and one he particularly liked - Sam had freely lent it to him for weeks at a time and Dave had taken some very nice photos with it.

Also in the box was a pre-WWII Colt .45 Single Action Army revolver in near mint condition. Sam had acquired it years before in a trade, and in the late 1930's had carried it on month-long hunting trips into the Coast Mountains, and had once shot a grizzly with it at about 20 feet - "Four shots in the head, faster'n I ever thought I could shoot!" Sam had told him, demonstrating how he had held the gun in his right hand and, closing his left hand over his right, arms rigid in front of him, cocked and dropped the hammer several times in rapid succession with his left thumb. The bear had gone down at the third shot, and Sam had sent the 4th slug smashing into the brain pan of the enormous skull at less than 8 feet.

The rest of the box could wait for another day. Here alone were two items that he would treasure for their own sake, and doubly so for their connections with his old friend. He took the Colt downstairs and cleaned it, even though it didn't need cleaning. Like Nora said, it helped to have something to do.

Chapter 17
SOMETHING MORE VALUABLE

Nora called him late that afternoon - Sam was vomiting blood again, and she had just called the inhalator. Sam died before either Dave or the inhalator arrived.

The days that followed were sad and difficult for both Nora and Dave, nor were they made easier by a confrontation that occurred a week after Sam's funeral.

Oscar had come to the house to get Sam's rifles. Dave happened to be there at the time, packing up some of the equipment in the shop.

Oscar had proceeded to take not only the several rifles, but all of Sam's shooting gear - his shooting box and spotting scope, and all the reloading equipment. The confrontation started when he picked up the Sharps Borchardt.

"That's not among the things you were to have, Oscar," Dave told him.

Oscar cut him off sharply. "Now look, you Johnny-come-lately: Sam told me months ago that I was to get his rifles. I shot with him for 16 years. You come along 5 years ago, and pretty soon you're taking him to shoots, and to the rifle range, but that doesn't entitle you to any of his rifles."

"Sam asked me to help him build that rifle, and it was to come into my hands when he passed on. Besides, the scope..."

"I was to get all of Sam's rifles," Oscar snarled. "Which reminds me, where is his #4 Lee Enfield?"

"Sam made me a personal gift of the #4 the day he and I shot at Fort Lewis in September. Nora will confirm that, and the matter of the Sharps also."

"What's that limping bitch know about it?" Oscar exploded, "Sam ..."

"HOLD IT!" Dave cut him off. "Don't ever make another remark like that. If you ever say anything like that again, I'll break your jaw on the spot. Without a word of a lie, I will. As for the Sharps, take it. But the sco...."

"That suits me just fine!" Oscar snapped.

"As I was going to say, the scope belongs to me." Dave took the Sharps Borchardt from the older man with a finality that brooked no further argument, and loosened the scope mount screws with a coin. When he finished, he set the scope on the workbench. "Let's go put this stuff in your car. You take the Sharps, and I'll carry some of the other things," he added, handing Matlin the rifle.

They found Nora in the front yard, looking at a newly erected "House For Sale" sign.

"What's this?" Oscar asked, surprised. "Are you going to move into an apartment, Nora?"

Nora took three steps over to where Dave stood, and slipped her arm around his waist. "Not an apartment. Dave and I are going to be married next month."

Dave smiled down at her, in time to see a look of surprise cross her face as she noticed the Sharps in Oscar's hand.

"What are you doing with that?" she asked.

"Well, I came to get Sam's rifles. You know they were to come to me when Sam died...," he said defensively.

"The Sharps was not one of the ones you were to have. Gramp..."

Dave stopped her gently. "It's all right, Nora. Let him take it. I'll explain to you later." To Oscar he added, "Oh, by the way, we never got around to bushing the firing pin."

"....I wasn't going to haggle with him over Sam's belongings. If he wants the Sharps that badly, he can have it. Besides, if I'd had a big row with him over it, it'd have reminded me of that every time I looked at it from then on - I'd never have enjoyed owning it. Oscar knew your grandfather a lot longer than I did, and yes, he ended up with his rifles. But I think that's about all he did get from him. I think that what I learned from Sam - like being able to let Oscar take the Sharps rather than fight with him over it - I think that's a lot more valuable."

"Grampa would have been proud of you for the way you did with Oscar over the Sharps, Dave. I am, too. And I'm glad you're not bitter about him taking it. But what'll you do with the barrel you ordered, and the wood for the stock?"

"Well, you've heard me and Sam speak of Sid Cooper, the gunsmith? I think Sid has a Ruger #1 Single Shot action he wants to sell. The Sharps would have been nice to have because of the association with Sam, but I've always favored the looks of the Ruger Single Shot, and mechanically and metallurgically it has to have it all over the Sharps. I think I'll buy that Ruger action off him, and build a rifle around it... and I'll color caseharden the buttplate Sam made, and use it, too."

"Grampa would like that. He'd say, 'Now there's a rifle that is the real Bullseye Mixture!'"

NOTE: **Barium Carbonate is a poison**, and must be handled *extremely carefully*. Further information on this fact, and the whole topic of backyard casehardening - truly "the final frosting on the cake" - is given at page 6-10 in **TMBR#3**.

ACKNOWLEDGEMENTS

The technical information on pack casehardening incorporated in this story was drawn from many sources. No one, so far as I know, has done what I have tried to do here, namely to set forth explicit and complete instructions on the whole matter of pack casehardening.

The fruit pit charcoal recipe is a case in point. This was reported in M.E. in the 1930's by someone who signed himself "VWDB". His instructions included the use of 2 oz. of barium carbonate, and 4 oz. of sodium carbonate, and said to mix this with "the charcoal". How much charcoal? That he didn't say, nor did he comment on the function of these two ingredients, or the benefit of their use. Those points, and others, were deduced from my own further reading of other published material in a number of different sources.

Three individuals who took the time to share their considerable/specialized knowledge with me were: Tom Bell, professor of chemistry, Simon Fraser University; Bob Buttons, professor of metallurgy, University of British Columbia; and Ben McKinnon, heat treating plant foreman, McLeod & Norquay Ltd Vancouver.

Another individual, John Fenwick-Jones, enthusiastically recommended the use of bone meal; when I first talked to him, he'd just made a new bolt for a P17 Enfield, and heat treated it thus. (Note: I do NOT recommend any reader of The Bullseye Mixture try making and heat treating such a highly critical part as a rifle bolt. Mr. F-J was trained as an armorer in the Canadian Army, and therefore has a better-than-average understanding of what he is doing in this area.)

Warpage is a potential problem in all heat treating, including pack casehardening, but it is less of a problem in pack casehardening than in some other forms of heat treating.

APPENDIX A

Concerning the matter of making division plates
from bandsaw blades (TMBR#1, page 44)

If your copy of **TMBR#1** predates the 8th printing, at page 44 you'll find an indented paragraph headed "FOOTNOTE TO THE ABOVE NOTE" wherein Mac says (erroneously) that even a fairly serious eccentricity in the plywood disk/lathe spindle axis setup will not have much effect on the accuracy of dividing.

One of my customers questioned the correctness of this statement, and when I bounced it back to Mac, he said "I don't know what (if anything) I could have been thinking about when I wrote that. Looks like I have dropped a clanger."

In the 8th printing of **TMBR#1** I edited Mac's comment to eliminate this erroneous statement.

A note re the Small Scribing Block shown at page 97, in TMBR#1

In the front view drawing of this item, at the bottom of page 99 of **TMBR#1**, please note that what might be mistaken for a loose washer between the tapered Scriber Arm and the Body, is in fact a flange, integral with the Spindle itself. Thus, you will likely make the Spindle from a piece of 1/2" CRS. Or make from 1/4" CRS, and silversolder or press fit the flange piece to same, and then machine it to spec.

NOTE: The **Handbook of Spring Design** (see page 154, in **TMBR#1**) now (1994) costs about $10.

RIFLE BARREL MAKING

After reading "The Secret of the Old Master", by Lucian Cary, in TMBR#1, several readers have asked me where more info on rifle barrel making could be found. Flip to page 204 of the book you are now holding, and in the upper left hand corner, you will see full info on **a 3 hour video I made in 1995 on how to make and use a rifling machine**. This video shows a machine any serious home shop machinist could make, if interested in this line of activity. **The machine will drill, ream and rifle a match grade barrel, starting from a solid bar**. The video is very thorough, and I have received photos from guys who have bought the video from me, and then built machines like the one shown in it. **(Also, don't overlook the book described immediately below the Rifling Machine Video!)**

Also, you might want to see the following articles, which appeared in other publications:

(1) "How Peterson Rifle Barrels Were Made", in *The American Rifleman Magazine*, Aug. 1960, page 26. This article is interesting, although it is about Peterson rather than Pope barrels, Peterson being a contemporary of H.M. Pope.
(2) "George Schoyen - Riflemaker Extraordinary" (Schoyen was another contemporary of Pope) at page 65 of the **1971 Gun Digest**. The photos of the interior of Schoyen's workshop show the machine tools he used, and will interest you. (See also the bibliography at the end of that article.)
(3) "Ramrod Guns, Country Style", in the **1968 Gun Digest**, page 204. and
(4) "They Made the Best Barrels", in the **1974 Gun Digest**, page 27. This article describes the work of Pope, Peterson, Schoyen and Zischang.

Making your own Chambering Reamers

A fella in Kentucky asked me for info on making chambering reamers. In the Sept. '73 issue of *The American Rifleman* (TAR) Magazine, at page 32, there was an article by J.R. Allington entitled "Make your own chambering reamer in just 2 hours". I believe this was for a D-bit type of chambering reamer. If your local library does not have back-issues of TAR, you can write to the NRA in Washington, DC and ask for a photocopy of the article.

The main problem that I would see for anyone trying to make reamers to SAAMI specs would be accurate duplication of tapers and radii. Fluting of reamers is well covered in *Machinery's Handbook* and similar books. Heat treatment raises the possibility of warpage. One technique is to chuck the reamer in a drill press, switch the drill press on, heat the spinning reamer with a torch and then lower it into the center of a can of water placed on the drill press table, the water in the can having been set swirling before quenching. Another method to beat the warping bugboo is to grind the reamer to D section (or grind the flutes into it) after hardening... ok on a small reamer such as for a bullet mold, or a bullet swaging die, but no small task for a whole chambering reamer. One could also send one's reamers to a heat treater such as the specialists used by knife makers.

Beyond that, the best I can say is, good luck.

Re: Gearcutting

I had planned to have a section in this book on the subject of gearcutting, but it did not work out. However, the people who produce *Model Engineer Magazine* have just recently published a little book on the subject, by Mr. Ivan Law. You might want to get it - I have only had a chance to look at it in a most cursory manner, so cannot say if it is real good, or just ok. It does have some neat stuff in it.

A bench gear cutting machine of quite comprehensive capabilities was designed by T.D. Jacobs, and described by him in an article in *M.E.*, starting January 16th, 1976. This is a hobbing machine which will cut just about any type of gear you want, short of a herringbone gear.

Castings and full working drawings (price not known) for Jacobs' machine are being marketed by the undernoted company. For details, write to: Preserved Technology Ltd., 4 The Shaws, Uppermill, Oldham OL3 6JX, England

We also sell drawings for some things you might find useful in your shop

The TINKER Tool & Cutter Grinding Jig: a simple, compact and effective jig, about as big as a basket ball, that teams up with your bench grinder to sharpen almost any cutter in the average small machine shop, **including end mills** and many others. You can build it from our detailed drawings.

Lautard's Octopus is a distant cousin to the Toolmaker's Block. It is a machined all over octagonal block of cast iron. If you've ever wanted **a small, multi-purpose work-holding fixture**, you should make **Lautard's OCTOPUS.**

Lautard's OCTOPUS incorporates some of the features of a bench block, V-block, vernier protractor, direct indexing fixture, tilting angle plate, rotary table, and layout jig. Ideal for those tricky jobs that come up every so often in your shop. Useful on your lathe H/S and T/S, mill, drill press, surface plate, and workbench.

The detailed working drawings include machining instructions, plus several "extra info goodies", like how to make a highly useful lathe faceplate accessory that'll hold ferrous and non-ferrous workpieces with a ferocious, non-marring grip.

A Universal Sleeve Clamp: Working drawings and instructions for making a highly useful accessory for your surface gage or dial indicator base. Our drawings include a worksheet on which you can scale the design up or down from the size shown, because we think you might want to - a larger version could be used as the basis for jointed-arm lamps for your workshop or living room.

A 3.75"Ø Rotary Table: (See photo at page 93 of **TMBR**) Full working drawings, plus detailed instructions that guide you through every step of what only looks like a difficult job. A "braggin' project" and a very worthwhile addition to your shop even if you already own a larger geared RT. (Could also be scaled up or down, as mentioned at page 145 herein.)

Send $1 for our current catalog of working drawings & instructions for the above items.

APPENDIX B
(A far-from complete listing of suppliers and magazines catering to
the interests of people who would be reading this book.)

Home Shop Machinist Magazine
and *Live Steam Magazine*
Traverse City, MI 49685
(800) 447-7367

Modeltec Magazine
P.O. Box 9
Avon, MN 56310
(320) 356 - 7255

Strictly IC - see page 146

Model Engineer Magazine
Nexus House
Boundary Way
Hemel Hempstead, Herts
England HP2 7ST

For *Model Engineer Magazine* subscriptions in
North America, contact **Wise Owl Publications**,
4314 West 238th St., Torrance, CA 90505-4509
(310) 375-6258

Engineering in Miniature
TEE Publishing
Edwards Center, Regent St.
Hinckley, Leicester
England LE10 0BB

The British Horological Journal
Upton Hall
Upton, Newark
Nottinghamshire
England NG23 5TE

Rifle Magazine
6471 Airpark Drive
Prescott, AZ 86301

Cole's Power Models, Inc.
P.O. Box 788
Ventura, CA 93001
(805) 643-7065

D&M Model Engineering
P.O. Box 400
Western Springs, IL 60558
(312) 246-3618

Metal Lathe Accessories
Box 88
Pine Grove Mills, PA 16868

Arrand Engineering
The Forge, Knossington
Nr. Oakham, Leics.
England LE15 8LN

Woking Precision Models
10 New Street
Oundle, Peterborough
England PE8 4EA

Neil Hemingway (See page 68, TMBR#1)
**We sell drawings for a number of
Hemingway's machine tool accessories**
Contact us for more information. GBL.

Gunsmithing Supplies:
Brownells Inc.
200 South Front St.
Montezuma, IA 50171
(641) 623-5401 (office)
(641) 623-4000 (orders)
www.brownells.com

Conover Wood Lathes
1-800-433-5221

Gilliom Manufacturing, Inc.
P.O. Box 1018
St. Charles, MO 63302
(kits for various shop tools)

J&L Industrial Supply Co.
(800) 521-9520, or (313) 458-7000

Production Tool Supply Co.
(800) 366-2600

Manhattan Supply Co. (MSC)
(800) 645-7270

Reid Tool Supply Co.
(800) 253-0421

Smithy Company
(800) 345-6342

Travers Tool Co. (TTC)
(800) 221-0270

KBC Tools (800) 521-1740

Enco Mfg: (800) 873-3626

Harbor Freight: (800) 423-2567

Blue Ridge Machine Tools (800) 872-6500

US Industrial Tool & Supply Company
(800) 521-7394
(a good source for sheet metal tools)

Carbide balls for "ball-sizing"
Spheric Inc., Suite 101A
428 West Harrison Street
Claremont, CA 91711

"KEEPBRYTE" - an anti-scaling
compound for use when heat
treating tools, etc. - get from
Travers Tool, etc.

Temperature indicating pellets:
Tempil Division
Big Three Industries Inc.
2901 Hamilton Blvd.
South Plainfield, NJ 07080

Lettering Guides
Pickett Industries
One River Road
Leeds, MA 01053
(413) 584-5446

AND OF COURSE, THERE'S US!

As you will find on the following pages, we offer numerous items not carried by those who retail our books. If you, or anyone you know who is a machinist or gunsmith, would like to see our full range, send us $2 for a paper catalog. Better still, go to our website at **www.lautard.com**. And of course, if you want an *autographed* book, there's only one way to get it: order direct from me, and ask that your book be autographed. Watch for notice of new items from us in *Home Hope Machinist Magazine*, and more so on our website. If you have reason to contact us (other than with an order*), an e-mail message or a phone call is usually better and faster than a letter. Please note that we are on West Coast time; wake-up calls are NOT appreciated - I often work late into the night, and get up later than I would if I had a real job. If you do write with a question, put the correct postage on your letter, and include a self-addressed envelope plus two international reply coupons.

　　　* **Please note**: while we do not accept credit cards, some customers have paid us with PayPal. If you want to make payment this way, you will have to investigate the matter for yourself on the Internet.

Note: I am also an authorized Myford agent, and can advise on, and arrange, purchase of a new Myford lathe or mill, or parts for existing machines. I have used Myford lathes for over 25 years, and I know these machines intimately.

Guy Lautard, 2570 Rosebery Avenue
West Vancouver, B.C. Canada V7V 2Z9
Phone: (604) 922-4909
website: www.lautard.com

Addition to the second and subsequent printings of TMBR#2:

More on Aluminum Castings

My friend Paul Scobel borrowed my copy of the article on castings recommended at page 149 herein, and after reading it, we had an interesting chat. Paul's view was that while the author's methods no doubt worked for him, he had a tendency to substitute mechanical fixtures for plain old "hands on experience"; or, to put that another way, he went to unnecessary lengths to get where he wanted to go.

Making cast aluminum flasks, for example, is not necessary, altho it will work. Ordinary plywood flasks would have served as well, says Paul. Mostly what I want to add to this 2nd printing of TMBR#2, I guess, is that while the article is good, and has merit, don't swallow it hook line and sinker without doing a little thinking on your own.

Paul also said that from his reading of Part 5 of John Pilznienski's 5-part serial "Home Shop Metal Melter", which ran in Home Shop Machinist Magazine in 1989, he felt that Pilznienski's approach was much closer to commercial practice, and that given time, money, and that article, one could turn out usable castings.

You might ask, "What does Paul know?". In his teens, Paul worked summers, weekends, etc in a foundry his father operated very successfully for many years in San Jose, CA. The Scobel foundry made a specialty of close tolerance, thin wall aluminum castings that other foundries could not duplicate. The Scobel family's ability to turn out such castings was based upon experience - of the materials used, the furnaces, the sand, etc, - rather than some magic touch that no one else had. When Paul talks about foundry work, I listen.

A customer commented upon my *'d footnote at the bottom of page 16 of this book, re whether or not the term "forged" was appropriately used by John McHenry in his article on MAKING AND CHECKING A FLAT SQUARE. The letter got read and filed, and now when I need it, I can't find it, so I can't credit the guy, but basically what he said was this:

"... Don't under-estimate the abilities of those old blacksmiths: for them to put a right angle bend in a piece of material, and then hammer it out flat, to produce the rough forging for a square such as McHenry described, would have been but a few minutes work, and not at all difficult...."

I can buy that. So maybe my footnote is just the mumbling of one from the younger generation, who never saw what those old fellas could do.

THE MACHINIST'S BEDSIDE READER
contains a wealth of know-how and info (some of it obscure, but all useful) that would take you countless hours to dig up on your own, PLUS..

Working drawings and detailed instructions for making 15 useful and practical machinists tools and lathe accessories.

Dozens of hints, tips, and tricks to help you get things done faster, easier and better in your shop.

A collection of about 2 dozen machine shop anecdotes.

2 highly readable machinists' short stories:

- The fascinating (true) account - and photos! - of a beautiful little lathe secretly built and used in a Japanese POW camp.
- "The Secret of the Old Master" by Lucian Cary. A young toolmaker attempts to learn the secret behind the legendary reputation of a master rifle barrel maker.

Partial Table of Contents

The Scrap Patrol
Standards of Workmanship
A Source of Welding Rod for Fine Welds
Filing in the Lathe and Elsewhere
Two Not-so-Common Ways of Measuring Holes
Some Notes on Reaming and Tapping
Drilling to an Exact Depth
How to Sharpen a Centerpunch (properly!)
How to Get More Work Done in the Shop
A Sharpening Jig for Drills from 1/8" to #60
A Small Depth Gauge and a Small Tap Wrench
Swivelling Base Fixture for a 2" Wilton Vise
How to Make Your Own Reamers
How to Make A Floating Arm Knurling Tool
A Tool for Straight Knurling
Fruit Acids and Fine Tools don't Mix
Holding Thin Work in the Chuck
A Graduated Handwheel for the Lathe Leadscrew
  (and a "Rubberdraulic" Slip-Ring Lock)
Ball Turning without Specialized Attachments, and
an Improved Type of Ball Handle
Oddleg Artistry
A Finger Plate
Designing and Fitting Split Cotters
A Small Scribing Block
Cutter Blocks and Shop Made Cutters
Some Info on Silver Soldering

A Toolroom Grade Sling Swivel Base for
  Tubular Magazine Rifles
Gun Making (and some tales from one who did)
Making Bullet Molds
The Ultimate Box Latch
Working Drawings for a Machinist's Toolchest
How to Make Imitation Ivory
Why not Build a Harmonograph?
A Set of Heavy Brass Napkin Rings
A Hand Beading Tool
Some Notes on Spring Making
3 Useful Accessory Faceplates for your lathe
Metal Polish - Bought and Home Made
A Corrosion Preventive Cutting Compound
A Fair Return for One's Work
A Chinese Tool Steel that Outperforms HSS
Good Advice on Getting Ahead
"Stealing the Trade"
Helping the War Effort
How Not to Get a Welding Ticket
Delphon and the Adding Machine
The Sleepy Apprentice Boy
A 19th Century Machinist's Apprenticeship
How to Impress Your Mother-in-Law
One Way to Ruin a Lathe
Removing a Jammed-on Chuck - the method of last resort
Dimensional Matrix for 6 sizes of Toolmakers Clamps

Comments from machinists who have read The Machinist's Bedside Reader

▪"Absolutely superb. Thoroughly enjoyable, with much knowledge." (D.S., Vallejo, CA)

▪"..an excellent book I would recommend to anyone even remotely interested in working with metal." (J.S., Custer, WI)

▪"..enjoyable and informative ... I love the manner in which the material is presented." (C.A.P., Jonesville, VA)

▪"Great. I teach machine shop, and the kids read it without being asked." (D.B., Ann Arbor, MI)

106 clear and detailed illustrations, 210 pages
8½ x 11", soft covers

OTHER PUBLICATIONS and PROJECTS from GUY LAUTARD

A VIDEO from Guy Lautard:
BUILD YOUR OWN RIFLING MACHINE

This 3 hour video shows full technical details on the construction of a rifle barrel making machine the basement machinist can build, and then use to produce match-quality cut rifled barrels. Watch also - often close up - as a barrel is drilled, reamed, and rifled from the solid bar. Thorough explanations are given at every stage - deep hole drill geometry, reamer details, making your own rifling heads, etc.

If you want to make your own rifle barrels, on a non-commercial basis, this is the way to do it. The machine shown in this video is in use today making match-winning 6mm benchrest barrels. Its owner/builder is a mechanical engineer who is also a gunsmith of some 40 years experience. This video will save you **endless** frustration and wasted hours. Just the part on sharpening your own deep hole drills will quickly save you the entire cost of the video. Comes with 36-page written supplement loaded with tips, drawings, info and suppliers' addresses.

Concludes with a 15-minute "short" showing several very fine old-style single shot rifles made entirely from scratch by a retired tool and die maker. This part will have you on the edge of your seat for sure!

THE J.M. PYNE STORIES AND OTHER SELECTED WRITINGS BY LUCIAN CARY

If you've ever read any of Lucian Cary's J.M. Pyne stories*, you'll want this book! Destined to become a classic and a collector's item, this handsome volume is a MUST HAVE book for machinists, gunsmiths and shooters. These wholesome and heartwarming fiction stories - ripe with the fragrance of gunpowder and cutting oil - combine humor, suspense, insights into human nature, and a certain amount of shop wisdom. They sprang from the pen of master story teller Lucain Cary more than 50 years ago, in part as a direct result of his long friendship with legendary rifle barrel maker H.M. Pope. Also included are several other of Cary's equally entertaining, instructive and uplifting stories and articles. You will read and re-read them all, with fresh enjoyment each time.

■ Almost a Gun Crank ■ H.M. Pope - Last of the Great Gunsmiths ■ The Rifle Crank ■ The Big Game Hunting of Rufus Peattie ■ Madman of Gaylord's Corner ■ The Old Man who Fixes the Guns ■ Forty Rod Gun ■ Johnny Gets His Gun ■ Center Shot to Win ■ J.M. Shoots Twice ■ The Secret of the Old Master* ■ No Choice ■ Revenge in Moderation ■ Harmless Old Man ■ Let the Gun Talk ■ The Guy Who had Everything ■ How's That? ■ I Shall Not Be Afraid.

* See TMBR#1, page 104. "No Choice" and "Revenge in Moderation" are Parts 2 and 3 of a trilogy, of which "The Secret.." was but Part 1. *6" x 9", 337 pages, quality printed, soft bound.*

TABLES & INSTRUCTIONS FOR BALL & RADIUS GENERATION

Part 1 of this handy little book shows you how to make ball handles, ball end mill blanks, etc. in your lathe, without special tooling - just a parting tool - by the 'incremental cut' method shown in somewhat less detail in TMBR#3. **Part 2** contains full instructions for generating male or female radii with a ball end mill, in a vertical mill. Say you want to machine a 1/4" radius on the corner of a piece of material. If you've got a 1/4, 5/16, 3/8, 1/2"ϕ or other size ball end mill you can do it easily, quickly and accurately. You won't believe your eyes, the first time you try it! **Part 3** consists of 70 tables (one per page) for the generation of both ball and radii from 1/64 to 1" radius.

On the inside back cover there's a memory aid you can flip to if you need to, then flip back to the table that suits your job, and go right to work.

Also includes notes and a photo showing how to rig an indicator on your lathe, which gives you direct number readout of carriage position when carving out a ball. The incremental cut system can be used to carry out some part-circle machining jobs, as well, and is also useful for making handles for feedscrew dials, levers, etc. This handy little book therefore includes some info, a drawing, and numbers for a very nicely proportioned feedscrew dial handle - if you make one, you'll fall in love with it the first time you use it!

Durable, soft covers, coil bound to lay flat in use; 104 pages, 4" x 6" to fit conveniently in your toolbox.

A BRIEF TREATISE on OILING MACHINE TOOLS

Here's the 'mother lode' on a topic of major importance to every machinist - 25 info-packed pages on ■ How to make a *really* good oil gun, in less than an hour, that will put oil where you want it, at 10,000 psi if need be! (it's a dream come true for Myford lathe owners!) ■ How to arrange for centralized oiling of machine tools using a one-shot oil pump, or the shop-made oil gun described above, how to remove spring ball oiling points, how to make and mount a centralized oiling manifold, what to use for oil lines, etc.. ■ Info on oiling machine tools - what types of oil are appropriate where, and why. ■ Complete working drawings for a handsome variable feed drip oil cup to fit lathes, mills, etc. Permits you to see that oil is definitely being fed, and shuts off tight so you don't waste oil when the machine is off. You'll like it. ■ Complete drawings for pea-sized drip feed oil cups for model work. ■ And more, including a recipe for a good leadscrew lube, and how to go about grinding in a lathe without harming it.

A UNIVERSAL SLEEVE CLAMP

An interesting exercise in machining, boring and screwcutting, the Universal Sleeve Clamp provides a clever means of hinging and clamping two rods together. It is the ideal accessory for your magnetic dial indicator base and/or surface gage, and can also be used as the hinge joint for a large capacity compass, or inside or outside caliper, much like a firm joint caliper. Scale up the design just a little and use it as the basis for a universal lamp for your lathe, mill or workbench. Made slightly larger again, you could use it to make a handsome and unusual living room lamp. (The drawings are arranged in such a way as to make designing new ones at any size an easy matter.) Also includes working drawings for a height adjustable tool holder for small boring tools and internal screwcutting jobs, plus a sheet showing the sort of boring tools I make. These incorporate a minor but significant feature not everybody thinks of when making tools of this type.

3.75"ϕ ROTARY TABLE

Even if you own a larger, geared rotary table, you will find this little ungeared Rotary Table (shown in **TMBR#1** at page 193) a most useful accessory for small jobs in your vertical milling machine: ■ mill a radiused end on a part, or round the ends of a set of loco conrods ■ space out holes accurately around a pitch circle ■ mill a circular slot ■ mill two or more slots at an angle to each other... and so on. If you make it at the sizes given in the drawings, the base will be 4" square, and the table will be 3.75"ϕ. Of course, you may want to scale it up, (or down, for the ultimate paperweight). Whatever size you make it, it's an interesting and instructive exercise in machining, and will be a prized addition to your shop tooling when done.

A LEVER FEED TAILSTOCK DRILLING ATTACHMENT

This item allows the drilling of small holes from the tailstock with ease, as it gives you a very sensitive feed to the drill, rather than the relatively insensitive feel you have when feeding a small drill forward by means of the tailstock barrel handwheel. This is not a major project, but it is one I think you'll like. It plugs into the tailstock barrel's taper socket; when you're not using it, you can whip it off the lathe and hang it on a nail on the wall.

A TILTING ANGLE PLATE (and more) FOR YOUR MILLING MACHINE

A tilting angle plate is a very useful addition to your milling machine. Most tilting angle plates - bought or shop-made - involve the use of castings. This one is made from aluminum plate, which eliminates the need for castings, and allows you to make it at the right size to suit your equipment. (An accessory of this sort would not last long in a commercial shop if made from aluminum, but it'll do fine in the hands of the guy who makes it for his own use.)

Also includes drawings for 5 other simple workholding aids for your milling machine: ■ a pair of locating/stop buttons ■ 2 types of stops for the front edge of the table ■ a "catch-the-corner" work stop ■ a set of small planer-style toe clamps, and ... ■ a big hold-down for use on a drill press or milling machine - this one's from an engineering professor who does "impossible" motorcycle repairs for his friends.

LAUTARD'S OCTOPUS

Part bench block, part V-block, part vernier protractor, part direct indexing fixture, part tilting angle plate, part rotary table, part layout jig, this multipurpose workholding fixture has more ways of holding a job than an octopus. Not for heavy work, but just right for small (and some not-so-small) and tricky jobs that come up every so often in every shop. Fully detailed drawings, machining instructions and how-to-use info are provided. Several 'extras' are included, like how to make a lathe sub-faceplate that thinks it's a magnetic chuck: if you have a piece of 3/8" or 1/2" aluminum plate handy, it'll cost you about $2 for the "power supply", and it'll hold non-ferrous as well as ferrous materials with a "ferro-cious" grip - this info alone is worth the price of the drawings! We can also supply an octagonal casting, if you can't scrounge up a suitable piece of steel.

A MICROMETER BORING HEAD FOR YOUR MILLING MACHINE

Based on a classic design by the well known George Thomas, this boring head's generously proportioned dovetail slide and dowelled gib, coupled with a large diameter graduated dial, make for precise and silky smooth adjustments; if well made, its repeatability is phenomenal. Takes 3/8"ϕ shank cutters, and, via adaptor sleeves, smaller ones as well.

You can use this boring head with simple shop-made cutters to machine washer seats, cut counterbores for socket head cap screws down to 0-80, bore holes for which you don't have a drill of the right size, and bore/counterbore holes in jobs too big to swing in your lathe. If you make it with a straight shank, you can mount it in a collet in your vertical mill, *and* in the 3-jaw chuck in your lathe. The drawings give all dimensions for making the Boring Head in 2 sizes, to suit both larger and smaller mills. This is a quality project that will serve you for a lifetime. The instructions are very detailed, to enable anyone to make this valuable and versatile milling machine accessory, *providing he is willing to take the necessary time.* You can be justifiably proud of yourself when it's done, and you'll learn a few things in the course of making it, too.

The TINKER TOOL & CUTTER GRINDING JIG

The TINKER is a simple, practical, and compact Tool and Cutter Grinding Jig that teams up with your existing bench grinder to sharpen end mills, side and face milling cutters, slitting saws, lathe toolbits, twist drills, reamers - in fact, just about any cutter you'd find in a small machine shop, a gunsmith's shop* or home workshop.

You can build the TINKER in your own shop from our detailed working drawings and instructions.

The basic TINKER consists of only 27 parts in all, most of them very simple to make. The drawings consist of over 60 pages, with many helpful notes throughout to make the project easier. They include notes for building, as well as complete user instructions. We can supply castings, or you may prefer to make your own welded substitutes for the castings, and that is easy enough to do.

NOTE: We want you to know what the TINKER will and will not do before you buy the plans and/or castings. Therefore, please send $1 for our 5 page illustrated brochure which provides full details about the TINKER Tool and Cutter Grinding Jig. Or ask for it free of charge if you are ordering other items from us at the same time.

* Not suitable for sharpening chambering reamers. Chambering reamers should be returned to the maker for sharpening... **but see TMBR#3 for a goldmine of info on how to hand stone your own chambering reamers!!**

The COLE DRILL

A versatile, hand-cranked drill that will do most things that your drill press can do, and quite a few things your drill press won't do, including such feats as drilling right through a 1" square HSS lathe toolbit! Broken engine block studs, wheel studs, farm, logging and mining machinery, and vehicle repairs, grain elevator maintenance work, boat building -- if you are involved with any of these types of activities or problems, you may find the Cole Drill very useful. For more details and ordering information, see TMBR#3, or contact us.

VERNIER PROTRACTOR

We now stock and sell very nice quality 2-minute vernier protractors made in mainland China. As one customer said, "Every other protractor I own was designed wrong in the first place." He likes this one a lot better.

CLOCK MAKING FOR MODEL ENGINEERS

We have a new book! Written by C.J. Thorne, a professional, full time clockmaker in England, and edited/pubished by Guy Lautard.

CLOCK PLANS

We also sell plans for 16 clockmakers tools, 7 clock movements, and 2 sundials, also from C.J. Thorne, of England. These are detailed working drawings with helpful notes, but no detailed how-to instructions. Anyone with the appropriate basic knowledge of clockmaking - or the above mentioned book on **Clockmaking for Model Engineers** - will be able to work from them. For more details, ask for our "clock catalog".

For our complete Catalog, showing all of the foregoing and more, send $2 to:

Guy Lautard
2570 Rosebery Avenue
West Vancouver, B.C. Canada V7V 2Z9

Better still, see **www.lautard.com**

Added to the 7ᵗʰ Printing, January 1999:

A NEW VIDEO from Guy Lautard:

EXAMINING A USED LATHE AND MILLING MACHINE
– A MACHINE TOOL REBUILDER SHOWS YOU HOW

This 3 hour video will show you how to examine a used lathe or milling machine you may be thinking of buying - and it **could save you a lot of carefully saved money!** Dennis Danich, a machinist and machine tool rebuilder with some 30 years experience, shows you what to look for in a preliminary examination of a lathe and milling machine, and the few tools you should take with you to carry out several simple but crucial tests. He further shows how to determine the accuracy of the machine's alignments, feedscrews, etc.

You'll also learn how to adjust and lubricate both types of machines, so this video will be useful to you whether you are considering purchase of a new or used machine, or just want to tune up your own machines.

Also on this video:

◈ An introduction to the art of scraping and frosting of machine tool surfaces; an explanation of how and why it is done, and a demonstration of several techniques, including power scraping.

◈ A demonstration of how to carry out grinding operations in the lathe without harming it.

◈ How to restore a bench oilstone to virtually as-new condition on a cast iron lapping plate. (Sharpening scrapers wears grooves in the surface of the combination stones typically used for this task. Eventually the stone will become virtually useless, unless the surface is dressed flat again. We show you how to do it.)

◈ An inexpensive surface grinder well suited to the needs of the home shop machinist, and an important warning about buying a used surface grinder. (A worn out surface grinder is no bargain at any price, even if it carries a famous maker's nameplate.)

◈ Various shop kinks, and several handy shop accessories you can make.

◈ An 1890's 6-foot-stroke Gray planer in action, machining a 3' straightedge casting.

◈ Besides the above, we'll also show you briefly a beautiful little Chinese-made 2-minute vernier protractor which we sell, show you how it works, and turn it this way and that so you can have a really good look at it from all sides.

AND.... at the end of the video, there's a 12-minute "short" about **a privately owned P40 Warhawk**, bought in 1946 for $50!! You get a detailed look inside the cockpit, and all around the aircraft. And when the owner starts it up for us - wait'll you see the smoke and flames roar from the exhaust stacks!!

For ordering information, see bottom right corner of page 205.

WE'RE ABOUT TO BRING OUT A NEW BOOK!!!!

The book **"Hey Tim, I Gotta Tell Ya...."** is now out of print. **However,** I'm currently working on a new book, which will be partly a reprint/revision of the best material from **"Hey Tim...."**, plus an equal or greater amount of new technical stuff, plus a reprint of my fiction story, **Strike While the Iron is Hot**, plus a new fiction story, **Rebirth of Morgan Ranch**.

Keep an eye on my website, at *www.lautard.com* for news of the publication of this new book, hopefully by April of 2001.